Modal Choice
and the Value of
Travel Time

Modal Choice and the Value of Travel Time

EDITED BY
IAN G. HEGGIE

CLARENDON PRESS · OXFORD
1976

Oxford University Press, Ely House, London W.1

OXFORD LONDON GLASGOW NEW YORK
TORONTO MELBOURNE WELLINGTON CAPE TOWN
IBADAN NAIROBI DAR ES SALAAM LUSAKA ADDIS ABABA
KUALA LUMPUR SINGAPORE JAKARTA HONG KONG TOKYO
DELHI BOMBAY CALCUTTA MADRAS KARACHI

ISBN 0 19 828404 7

© Oxford University Press 1976

© Introduction, selection, and editorial matter Ian Heggie 1976

© each chapter the author of that chapter 1976

Printed in Great Britain by
Fletcher & Son Ltd.,
Norwich.

Contents

List of Figures

Introduction

by Ian G. Heggie

This book brings together a number of recent research studies which deal
with the dual topics of modal choice (by individuals) and the value of
savings in travel time. Both play an important part in the formulation,
as well as in the evaluation, of transport policies and plans. The former,
i.e. our assumptions about modal choice behaviour, have not only
influenced the physical planning process by implying certain patterns
of transport usage, but have also played an important role in the formulation of policies of traffic restraint and in the attempts currently being
made to encourage the usage of public transport.

As a by-product, most theories of modal choice provide a value for
what is euphemistically called 'the implied value of travel time savings'.
Savings in travel time are a common feature of most transport improvements, and the generally high values of time implied by most modal
choice models have turned them into the most important tangible
output produced by any transport investment. Although the values
assigned to these savings have declined in recent years (from 31 per cent
of the gross wage in the case of commuters in 1962 to 17 per cent in
1974) they are still of overriding importance and frequently account
for half or more of the estimated annual benefits (in the M1 and
Victoria Line studies, time savings accounted for 89 per cent and
80 per cent respectively of the benefits).

By emphasizing the importance of time savings the common assumptions about modal choice behaviour have seriously affected our entire
planning philosophy. They have done so in at least four ways:

1. The view that travellers want speed and are willing to pay large sums
   of money to acquire it has automatically downgraded public
   transport   the 'slow and clumsy' mode   as a serious planning solution.
2. Any scheme that speeds up traffic immediately shows a high rate of
   return over cost and appears 'justified'.
3. The importance of speed has made it the most important single
   engineering design parameter. Nearly all transport improvements
   attempt to speed up the traffic.

4. As a result of the emphasis on speed, transport facilities have become increasingly specialized. Indeed it is through specialization alone that improved speeds are often achieved. For example, roads that serve multiple purposes and allow free pedestrian movement, kerb-side parking, and slow traffic movement are often 'improved' by banning parking and containing pedestrians by means of iron guard rails.

The absurdity of the above process has fortunately provoked a common-sense response. Roads would continue to be built—but to lower standards—and the use of private cars would be restrained, while the use of public transport would be encouraged. Where possible, pedestrians would also be given precedence.

But how could the new grand scheme be implemented to ensure that it was fair and efficient? The theoretical foundations of the transport planning process remained unshaken. Schemes were thus designed, and implemented, utilizing the standard assumptions underlying modal choice behaviour, and it is only now—with several years of practical experience behind us—that we are able to look back and judge how satisfactory they are. It is part of the purpose of this book to do that.

Several books, and a large number of articles, have recently dealt with the joint topics of modal choice and the value of travel time. Why should this book be different, and what new insights can it offer? Its publication is clearly timely, in that the basic idea of designing for the free flow of traffic has been rejected and the whole transport planning process is being reorientated and pointed in a new direction. A reassessment of the assumptions and logical consistency of the basic concepts of transport planning is therefore overdue. But, although this reason was instrumental in prompting the preparation of this book *now*, it does not constitute its main—or even its most important—rationale.

Almost without exception, all recent research on modal choice and travel time has utilized a conceptual framework that severely constrained the results. Indeed, the framework—perhaps unwittingly—embodied assumptions about the causal and behavioural processes that the research itself was seeking to define. The results were thus conditioned from the start and could do little more than validate, albeit rather weakly, the presumptions already implied in the explanatory models. The most influential presumptions, which are questioned further in the research studies presented in this book, are as follows:

1. That human behaviour can be represented by continuous, linear functions. Jennings and Sharpe (Ch. 5) show in quantitative terms how this assumption has affected previous results, and demonstrate that the existence of thresholds (or of non-linearities) in the way that people value small travel time savings very quickly reduces the over-all level of benefits conventionally calculated. Heggie (Ch. 1) has likewise

discovered a consistent pattern of such thresholds; while Heraty (Ch. 3) suggests that thresholds might explain the very large discrepancy reported in her work between observed and predicted modal choice behaviour.

2. That individuals respond continuously to changes in objective information (excluding any threshold effects). In other words they do not exhibit habitual behaviour. Both Earp, Hall, and McDonald (Ch. 2) and Hensher (Ch. 4) question this. The former show that many travellers have little knowledge of alternative modes of transport (they do not acquire information about them as part of the choice process), and that altering relative modal attributes does not seem to affect patronage; while Hensher suggests that travellers only consider alternatives when their chosen mode deteriorates.

3. That people choose between competing modes by comparing the relative time and cost of making the journey in each alternative way, and adopt the one that minimizes the generalized cost (weighted time plus cost) of making the journey. Heggie (Ch. 1) not only shows that individuals exaggerate the characteristics of alternative modes (ex post justification) but that questions on car costs have been mistakenly interpreted to mean that car users perceive car costs in terms of marginal running costs when they do nothing of the sort. Earp, Hall, and McDonald (Ch. 2) likewise show that most travellers have little, if any, knowledge of alternative modes and could not possibly be engaging in any mental trade-off calculations.

4. That the decision to make a journey (trip generation) precedes, and is independent of, any modal choice decision. Changes in modal attributes only affect modal choice and do not lead to more or fewer journeys being made. Earp, Hall, and McDonald (Ch. 2) show that this assumption is invalid. A significant number of respondents in their inquiries indicated that they would have cancelled or postponed the journey had their chosen mode not been operating.

5. That the savings in travel time which accrue to people travelling on business are correctly valued at the gross wage rate, often inflated to allow for overheads and other direct employee expenses. Carruthers and Hensher (Ch. 6) show this to be incorrect. Business travellers not only do some of their travelling out of office hours, they also work while travelling and this is often more productive than office time because there are fewer distractions.

The present studies embody many other searching queries about the accepted views on modal choice behaviour and the value of travel time savings. In some cases they merely demonstrate that some of our existing assumptions are wrong; in others they suggest improved assumptions or try, in a more fundamental way, to expose the mental processes

underlying the way in which people engage in travel decisions or value
different modal attributes. Certain firm conclusions emerge, some of
immediate importance for policy. It seems fairly clear that:

(a) The modal choice process is not based on a mental comparison of
the attributes of competing modes, so that attempts to 'attract' users on
to public transport are probably doomed to failure. Public transport
patronage will only be significantly improved if the decision is forced by
means of extensive publicity or car restraint.

(b) Changing modal attributes can interact with the decision of whether
or not to make a journey. Policies like traffic restraint are therefore
likely to lead to fewer journeys rather than the re-allocation of existing
journeys to other (usually public transport) modes.

   A more fundamental conclusion relates to our whole travel choice
philosophy. It is evident from these studies that our current theories
of modal choice and the accepted values of travel time used in applied
planning studies are seriously at fault. In the long term we have simply
got to develop planning and policy models with a rigorously justified
behavioural basis. The term 'behavioural' must likewise be made to mean
what it says, i.e. that the structure and mathematical formulation of
the models reasonably represent the decision processes involved. Some
modal choice models use the term in an inappropriate sense. All they
do is model individual—rather than aggregate or zonal—behaviour in
terms of the standard quasi-behavioural assumptions being questioned
in this book. They also make the additional behavioural assumption
that modal choice decisions are made by individuals rather than by
sub-groups within an extended family unit.

   It is often argued that rigorous behavioural models would be too
complicated to be of any practical value in an applied planning context.
This is probably not true. The existence of thresholds, of habitual
behaviour, and of a lack of emphasis on the characteristics of alternative modes implies that the choice process—for modal choice decisions
at least—may be quite robust and that the new generation of behavioural
models might be marginally simpler. One thing is clear. They will
certainly be more closely related to our intuitive notions of how people
behave.

1

A Diagnostic Survey of Urban Journey-to-work: Behaviour *

by IAN G. HEGGIE

1.1. INTRODUCTION

The principal objective of this survey was to try and establish how people perceived and valued savings in travel time. It also collected, through a number of supplementary questions, general information on modal choice.

Savings in travel time are still the most important benefits generated by urban transport improvements, and yet there is still no satisfactory micro-economic 'theory of the allocation of time' that has been validated by real-world experiment. The theory of modal choice behaviour is better developed. However, even here fundamental issues of interest to the policy maker remain unresolved. Existing models have an extremely flimsy behavioural basis which fails to recognize the numerous metaphysical and perceptual assumptions (some of which can only be justified in special circumstances) underlying their structure and the methods of verification.

The purpose of the present survey was thus to help develop a more consistent theory of travel time valuation which could be reconciled with the established principles of consumer choice.

1.2. THEORETICAL FRAMEWORK

1.2.1. TRAVEL CHOICE AND TRAVEL TIME VALUATION

There are numerous ways of characterizing travel choices. However, since the present survey was concerned with establishing how people perceive and value travel time savings, it naturally focussed on the type of travel choice models which were capable of yielding a so-called 'value of time'. The principal concern was thus with those models—mostly disaggregate behavioural models—which characterize journey decisions in terms of a cost and time trade-off enabling the 'price' of time to be

* This survey was carried out in Vancouver in 1972/3 and was sponsored by the Transportation Centre at the University of British Columbia.

calculated from the relationships between the co-efficients.

The types of travel choice relevant to the present analysis were therefore:

(a) Modal choice decisions, in which behaviour is characterized in terms of an explicit trade-off between the relative times and costs of two competing modes. The functional relationship between the choice and the time and cost of the competing modes takes various forms and usually includes other socio-economic and household characteristics, such as income, car ownership, use of car for work, percentage of travellers who are male etc. (Refs. 1, 2, and 3).

(b) Route choice decisions, in which behaviour is characterized in much the same way as in modal choice models. The main difference is that route choice models generally concentrate on a more limited set of socio-economic and household characteristics, since car ownership and some of the other variables included in the modal choice models are clearly not relevant to route choice. (4.)

(c) Response to parking charges. This can be thought of as a sub-set of (a) with one important difference. Modal choice models collect data about the revealed preferences of people and seek to explain them in terms of a cost-time trade-off, qualified by the effects of certain other parameters. The parking charge model seeks to discover what increase in charges would just persuade people to change from their preferred mode to the next best alternative (5, 6), and usually does so by examining their stated preferences.

Research workers using these three approaches have sought to interpret their results as portraying the money value people attach to saving travel time in different travel choice situations. The theoretical background to this interpretation is nevertheless based on the flimsiest of foundations and this clearly detracts from the quality of their results. However, before we can attempt to present a possible framework for explaining the role and relevance of savings in travel time, a more fundamental question requires attention. Traffic engineering terminology distinguishes between trip generation, modal choice (in some models this occurs after trip distribution), trip distribution, and route assignment. These correspond to the parts of the travel decision which pose the questions: should I make a journey, what mode of transport should I use, where should I go and, if there is more than one way of getting there (route), which route should I choose? But it is assumed that each is an independent stage in a sequential trip-making decision. The only formal interdependence is reflected in common parameters, e.g. some trip generation models and modal choice models use car ownership as an explanatory variable, and in the iterative way in which estimated journey characteristics (e.g. link speeds) are manipulated until equilibrium values are reached. There is no interdependence in the actual process of making the decision. It is implicitly assumed that each component of the decision is independent and that one can analyse modal choices without reference to any other part of the journey decision.

Both our general knowledge of trip-making behaviour and the insights gained from other aspects of consumer bahaviour suggest that the sequential assumption is incorrect. The decision of 'whether' is intimately linked to those of 'how' and 'where' and cannot readily be separated except under special conditions. For example, in the case of shopping journeys, a possible increase in parking charges will not only affect modal choice, it will also affect *where* the shopping is done and *whether* the journey is made at all. A respondent can therefore answer a question like 'what parking charge would just persuade you to use a bus for this journey', knowing very well that the stated parking charge would actually lead to quite a different type of travel response. It might lead to fewer journeys or to the choice of a different destination.

The only obvious case in which this problem can be ignored is that of the journey to work. Since—at least in the short run—the decision to make the journey, as well as that of where to go, is effectively fixed; the traveller can only choose how and on what route to make it. With this exception, however, unless it can be shown that, contrary to inferential evidence, people do engage in journey decisions in a sequential way—or that the sequential assumption does not lead to any significant bias—analyses of travel choices based on such an assumption are likely to give a biased estimate of the value of time.

1.2.2. A MODEL OF TRAVEL TIME BEHAVIOUR

This model is an extension of an earlier model (7) which rests on one simple proposition: individuals perceive journey characteristics in several dimensions and their preferences between any pair of dimensions can be characterized by an indifference curve. Since the present analysis is primarily concerned with time and cost the proposition thus reads: if all other things are equal, a time-cost preference function exists for each individual making a specific journey by a given mode, relating the combinations of journey time and cost between which he is indifferent. If these functions are linear and parallel they will look something like the straight lines shown in Fig. 1.1.

Each line in the diagram represents the mode-specific combinations of time and cost between which the traveller is indifferent. However, although the two lines representing modes 1 and 2 are linear and parallel, there is no particular reason why empirically estimated functions should necessarily conform to this shape. Indeed, there are good reasons to suppose that preference functions are non-linear, and that, for leisure journeys, they might be concave to the horizontal (i.e. some

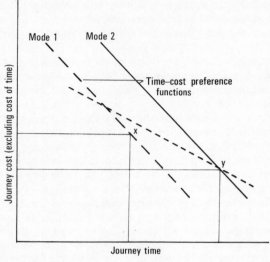

Fig. 1.1

positive utility is derived from the journey). Thomas and Thompson (4) argue that above points X and Y the preference curve is probably S-shaped: 'For very small amounts of time saved, empirical evidence indicates that motorists are insensitive to reductions in trip time, while economic theory suggests an eventual diminishing marginal utility of time saved as the amount of time saved continues to increase'. They could, of course, have argued differently. As the size of the time saving increases, the range of substitute activities available to the individual also increases: 'A one-minute saving on a work journey (which might not even be perceived) permits little substitution apart from an extra minute in bed or an extra minute reading the newspaper. A 20-minute saving, on the other hand, permits a whole range of substitute activities, from writing a letter to taking the dog for a walk.'(7.) So the marginal value an individual places on the time saving might increase as the size of the saving gets bigger. However, the shape of the preference function will also be affected by the nature of the journey (i.e. work or pleasure), by the marginal utility of time saved (i.e. beyond a certain point its marginal value might start to decline), by the effect of budget constraints, and by the effect—if any—of the reduction of over-all trip length. The combined effect of these factors does imply a non-linear relationship.

The above preference functions, however, do not measure the 'pure value of time'. They show its net uncompensated value: the amount of money people are just willing to pay to enable them to exchange different amounts of time spent on a particular journey for an alternative time-consuming activity. They thus represent the difference between the uncompensated opportunity cost of time—which can be loosely interpreted as the pure value of time—and the positive utility or disutility of travelling, under given circumstances, on the subject mode. The preference functions are 'uncompensated' because they assume that the marginal utility of money remains constant.

There is one further point of some importance. The opportunity cost of time wasted is the benefit foregone by not using it in its next best use. This will clearly differ between persons, for the same person in different circumstances (e.g. a five-minute saving on the journey to work generally provides a more limited range of substitute activities than a five-minute saving on the journey from work), the size of the time saving, and so on. So that, although there may be a pure value of time for a single individual under a given set of circumstances, this value is likely to vary between individuals and circumstances. An individual may therefore have a pure value of commuting time (if we ignore trip direction) or a pure value of convenience shopping time, but they will not necessarily be equal. The circumstances under which time is saved (or lost) are so different that the substitute activities defining the foregone opportunities will almost certainly differ and affect the individual's relative time values. Although the pure value of time may therefore be a useful theoretical concept, it is not likely to simplify the procedures used for empirically estimating the net values relevant to modal choice and physical planning decisions. The principal difference between time-cost preference functions and the usual indifference curves of micro-economics is that the traveller reaches a higher level of indifference by being on a lower curve and there is no particular reason why the curves should not intersect.

The first characteristic is quite obvious. The traveller is trying to minimize rather than maximize the journey characteristics measured on the two axes. The closer his final choice is to the origin the greater will be his consequent satisfaction: mode 1 is thus preferable to mode 2 in this diagram. The rule may nevertheless break down near the origin, since some individuals may derive positive utility from a short journey (e.g. they may view part of the journey as an essential transition from one activity to another).

The second characteristic is less obvious. However, if 1 and 2 represent a rail and a bus mode, it is quite conceivable that an increase in the speed of the bus mode (which might very well be accompanied by a rougher ride, especially for standing passengers) might be valued less highly than an equivalent increase in the speed of the rail mode. The preference curve for 1 might thus intersect that for 2 implying that each would be preferred over a certain range of speeds. Preference curves, being incompletely specified, cannot therefore be equated with indifference curves.

The standard modal choice models can now be illustrated with reference to these simple concepts. The most popular (2) utilizes the concept of the disutility of travel to explain modal choice. Each competing mode is characterized by a number of perceived dimensions (1 to k), representing characteristics like travel time, walking time, cost, etc. which are assumed to generate disutility. The dimension is called d and is subscripted with a p (1 to k) to represent the pth dimension, an i (1 to h) to represent mode i, and a j (1 to n_i) to represent person j on mode i. The dimension is thus written as d_{pij}, while its contribution towards the traveller's disutility is denoted by a weight λ_{pij}. The disutility of the pth dimension of mode i for person j is thus $\lambda_{pij} d_{pij}$ and his total disutility of travel by that mode is $D_{ij} = \Sigma_{p=1}^{k} \lambda_{pij} d_{pij}$, assuming that the contribution of each dimension remains linear. The traveller will then choose a mode i from amongst a set of competing modes for which $D_{ij} < D_{1j}, D_{2j}, \ldots D_{nj}$, i.e. he will choose the mode that generated the least personal disutility.

In the two-mode case (i.e. that of binary choice), when the distinction between walking, waiting, and in-vehicle time is ignored, the dimensions of disutility can be characterized as the absolute travel cost c and travel time t. This gives rise to the following two equations:

$$D_{1j} = \lambda_{11j} C_1 + \lambda_{21j} t_1$$
$$D_{2j} = \lambda_{12j} C_2 + \lambda_{22j} t_2$$

The above equations measure time and cost in absolute terms, although differences, ratios, a logarithm of the ratio, or some other relative measure are also sometimes used.

For purposes of empirical estimation the weighting factors, λ, are usually assumed equal for each individual *and* for each mode, i.e. $\lambda_{ij} = \lambda$. The relative disutility between modes thus becomes $R_{12} = D_{1j} - D_{2j} = \lambda_1(c_1 - c_2) + \lambda_2(t_1 - t_2)$. The individual will then choose mode 1 when $R_{12} < O$ and mode 2 when $R_{12} > O$. The implied value of travel time savings is simply the ratio of the coefficients for time and cost, λ_2/λ_1.

The above models are usually estimated by means of discriminant analysis. Stated values for c_1, c_2, t_1 and t_2 are fed into the model and the weights λ_1 and λ_2 are then estimated by choosing values which minimize the number of users of modes 1 and 2 who are 'misclassified' by the model. In practice this corresponds to finding the linear indifference curve which maximizes the number of 'preferred' choices lying to the left of the curve and the number of 'rejected' choices lying to the right.

The theoretical principles underlying this estimation procedure would probably be valid if the modal choice population was homogeneous. Unfortunately it is not. There are several groups of travellers who exhibit quite different patterns of behaviour. Two principal groups differ in the following respects: the first group pay *less* money but spend *more* time making the journey; the second group pay *more* money but spend *less* time making the journey. There are clearly other groups as well, i.e. those who both pay more and spend more time making the journey, but they are usually excluded from the analysis on the ground that they exhibit 'irrational' behaviour! Their exclusion is invalid, but it will nevertheless be assumed for purposes of the present argument that there are only two categories of traveller. Their behaviour is illustrated in Figures 1.2 and 1.3. Figure 1.2 shows how the first group behaves and shows how its indifference curve passes through point b so that point a, representing the rejected alternative, lies to the right of the curve. Figure 1.3 does the same for the second group, although in this case it is point a which lies on the indifference curve and point b which now lies to the right.

The slope of the indifference curve, which is equal to the marginal value of travel time, is λ_2/λ_1. However, it is clear from Figures 1.2 and 1.3 that the slope in Figure 1.2 will nearly always be different from that in Figure 1.3; the assumption that $\lambda_{ij} = \lambda$ (i.e. that the value of λ is the same for all modes and for all individuals) is thus violated and the resultant value of λ_2/λ_1 will be averaged over both groups. This clearly introduces a bias into the estimation procedure dependent on the proportion of

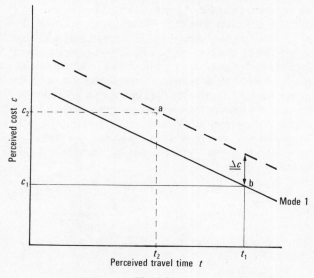

Fig. 1.2

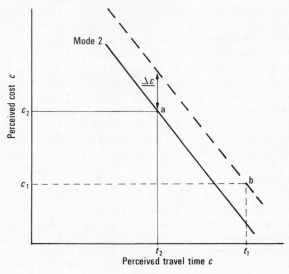

Fig. 1.3

each group present in the sample. When more than two groups are present, the estimate of λ_2/λ_1 will clearly be even more biased. The usual method of estimating a modal choice model does not therefore provide an unbiased estimate of the value of travel time.

To avoid this difficulty an alternative approach, proposed by Lee and Dalvi (5,6), makes use of the concept of 'diversion charges'. This measures the amount by which the cost of the individual's preferred mode would have to increase to just persuade him to switch to an alternative mode. In their empirical analysis they measure responsiveness to *increases* in cost, although they point out that they could equally well have 'diverted' users by means of a decrease in cost or by changing relative travel times. Each method, as they clearly recognize, might give slightly different results. If the diversion charge is equal to Δc, and the marginal value of travel time savings is constant and equal to v, the value of v can then be calculated as follows:

$$\text{Group 1:} \qquad c_1 + \Delta c + v_1 t_1 = c_2 + v_1 t_2$$

$$\text{(mode 1 preferred)} \qquad v_1 = \frac{c_2 - (c_1 + \Delta c)}{(t_1 - t_2)}$$

$$\text{Group 2:} \qquad c_1 + v_2 t_1 = c_2 + \Delta c + v_2 t_2$$

$$\text{(mode 2 preferred)} \qquad v_2 = \frac{c_2 - (c_1 - \Delta c)}{(t_1 - t_2)}$$

The value of v_2 might be expected to exceed v_1 and this is in fact confirmed in Lee and Dalvi's empirical investigation, which gives a value for v_2 nearly three times that of v_1. The available evidence therefore shows that the time-cost preference functions for the two modes have different slopes, i.e. they are not parallel.

It is worth emphasizing that diversion charges do not 'avoid' all the difficulties associated with discriminant analysis; they simply make the differences in group behaviour explicit by calculating separate values of λ_2/λ_1 for each group. They still leave several important questions unanswered. Apart from any queries concerning the precise way in which Lee and Dalvi estimated the slope of the time-cost preference functions, two major uncertainties remain:

(i) Are the functions linear, or nearly so over the range of observations involved?

(ii) Are they smooth and continuous, or do they exhibit thresholds?

There are also a number of other minor queries relating to the effect of journey length, whether the individual is travelling *to* or *from* work, how his disutility is affected by the level of congestion experienced, and so on. The prime objective of the present investigation was nevertheless to test for linearity and for the existence of thresholds.

1.2.3. METAPHYSICAL PROBLEMS OF ESTIMATION

The information used in estimating the above models is generally compiled from observations based either on 'hard' synthetic data, or on a set of characteristics supplied by the respondent, or sometimes both. A third procedure, which is only rarely used, is the use of uncontrolled experiments (e.g. one parameter is varied while the rest are held constant, and the result is monitored).

The use of hard synthetic data (e.g. measured travel times and costs) raises a number of philosophical problems. The subject phenomenon in each case is human behaviour: how do people perceive travel characteristics and respond to changes in them? The relevant data are thus the perceptions of the individuals involved, no matter how biased they may be. It is their perception of characteristics, or of changes in them, that leads to a behavioural response—not changes in the underlying hard data, of which the respondent may not even be aware. Models which incorporate hard synthetic data can therefore only be valid, in demonstrating a relationship between cause and effect, if they can explain how individual responses are related to real-world phenomena through an intermediate model of perception.

The alternative approach is to allow the respondent to supply his own perceptions of the travel characteristics being studied. This is an increasingly popular mode of data collection, but it suffers from its own peculiar set of problems. There are three main objections:

(a) The way in which questions are asked, and the provisions made for responding to them, often lead to a significant bias. In the discussion which follows this effect is termed 'experimental interference'.

(b) The individual's recorded perceptions may include an element of ex post exaggeration to justify his present pattern of behaviour.

(c) There may be a large gap between what people say they will do when asked e.g. 'how much are you willing to pay?' and their actual response when faced with a real-world choice.

Some of the above difficulties can be avoided by good questionnaire design, or by soliciting information in a carefully planned way. However, it will be argued in a later section of this chapter that an important form of 'experimental interference' has been allowed to creep into travel surveys and has contributed significantly to the general weakness of travel choice models.

The question of ex post exaggeration raises a number of difficulties. If respondents recorded the *actual* perceptions on which they based their decisions, no matter how exaggerated they were, it would not result in any particular bias in a behavioural model. However, if people have a set of implicit perceptions upon which they base their decisions, together with a set of stated perceptions which they then use to 'justify' these decisions, a very real source of bias is introduced.

The single most important objection to using stated responses (e.g. what level of parking charges would just persuade you to change modes) is that there is no guarantee that what the respondent says he will do will correspond to what he actually does when faced with the actual decision. This is in addition to any objection about the bias implicit in the way the questions are asked (i.e. they might relate to decisions beyond the range of the respondent's normal experience). This objection is nevertheless not as serious as it seems, although it does highlight the need to distinguish, for both analytical as well as planning purposes, between stated and declared values.* The reason is quite simply that neither is clearly 'right' or 'wrong'; each is relevant for certain types of planning and policy decision.

The argument runs roughly as follows. Certain planning decisions, for example the improvement of a certain stretch of road, give rise to very few externalities. The users enjoy most of the benefits and, by and large, because the externalities (e.g. noise, visual intrusion, etc.) are limited they also bear most of the costs (ignoring the way in which the road taxation system actually works). Non-users therefore have little or no interest in the specific benefits or costs. Since users are dominant, it is thus the declared value of time, rather than its stated value, that is relevant to the evaluation.

* A *stated* value refers to what a person *says* he is willing to do; a *declared* value to what he actually *does* when faced with a real-world choice.

At the other extreme one may have a project, for example a new airport, with very important externalities. The users may still be the most important single group of beneficiaries, but the costs, which could include substantial noise nuisance, relocation of homes, destruction of the countryside, etc., may have to be borne by the community at large. Now clearly, in such cases the community at large is usually entitled— within a democratic planning framework—to display the 'prices' at which it is willing to see the non-monetary items in the evaluation (e.g. the time savings) traded against the tangible monetary ones like savings in fuel costs. This is clearly a case in which people's stated values are going to be more important than their declared ones.

Both values therefore have some role to play in applied planning and policy decisions, although this role clearly depends upon the type of decision involved and on the incidence of the user and non-user benefits/costs. It is incorrect to assume that stated values are somehow fallacious because they do not relate to *actual* behaviour, while declared values are more reliable. Of course in one sense they are but then declared behaviour suffers from its own epistemological problems. We observe and measure declared behaviour, not the mental processes behind it. Any explanatory model of behaviour based on declared data must therefore impute motives and objectives to the individuals involved (i.e. the respondents *seem* to be minimizing travel time, or the amount of walking time involved but we do not know that they are). Models based on declared behaviour are thus on equally shaky ground and there is no logical reason to suppose that a model based on declared behaviour— but imputed motives— has any greater claim to certainty than a model based on stated behaviour which allows closer contact with individual attitudes and beliefs.

1.3. SURVEY OF JOURNEY-TO-WORK BEHAVIOUR

1.3.1. OBJECTIVES

The following diagnostic survey set out to explore three main aspects of urban travel behaviour:
(a) Is there a consistent bias in the way respondents describe the travel time characteristics of their preferred and alternative modes?
(b) Is there any form of bias in the way respondents describe the percieved cost of using a car?
(c) How do respondents value savings in travel time when these savings do not require a change of mode?

The first area of inquiry represents a partial follow-up to Quarmby's result (2):

. . . There was remarkable agreement between car users and bus users on the speed and time of car travel . . . But bus travel in the perception of

car users is about 20 per cent slower than that of bus users . . . it seems
that about half [this difference] is attributable to a difference in the
walking and waiting times . . . But half the difference occurs on 'in
vehicle' time, implying that there is a genuine difference in perception
between car users and bus users of about 10 per cent in the actual speed
of bus travel.

A consistent bias like this has serious implications for a modal choice
model utilizing the stated characteristics of competing transport modes.
The survey described here therefore examined whether car users consistently exaggerated the characteristics of the alternative bus mode. This
was done by collecting and matching the stated characteristics of the
bus mode perceived by bus users, against the same characteristics perceived by those car users for whom 'bus' represented the best alternative
form of travel. The same information was not collected for the car
mode, i.e. how both bus and car users perceived its characteristics, for
fear of unduly complicating the questionnaire and reducing the
response rate.

The second area of inquiry was dealt with by asking car users a series
of questions about car costs, although the prime objective was not to
discover *what* they said, but *how* they said it. This meant that each questionnaire had to be examined visually to see whether the way in which
the questions were answered indicated how the respondents arrived at
(or rationalized) their answers.

The final area of inquiry was the most important. It sought to find out
how people traded off travel time against money costs in situations in
which the time saving did not involve a change of mode. The aim was to
try and trace out the aggregate time-cost preference functions for a group
of individuals and to test to what extent they were influenced by either
journey or individual characteristics. The factors used as independent
variables in the analysis were:

(1) the size of the time saving;
(2) the over-all length of journey and the length of activity engaged in;
(3) whether the journey was being made *to* or *from* work;
(4) the geographical area in which the respondent lived;
(5) the income level of the respondent (only available for part of the
    sample);
(6) the relation between the responses for walking and waiting time and
    (a) the season of the year and (b) the provision of all-weather
    protection at bus stops.

For purposes of data collection, the journey was divided into walking
time, time spent waiting to change buses (this is an important aspect of
North American bus operations), and in-vehicle time. Although information was collected about the time spent waiting at bus stops at the

start of each journey, no attempt was made to develop time-cost preference functions for this part of the journey. The reasons for this are two-fold: average waiting times in Vancouver are very short (95 per cent of travellers wait for less than 10 minutes) and, second, buses in the city do not stray very far from their published schedules. Most waiting time therefore represents the respondents' own margin of safety.

1.3.2 SURVEY DESIGN

The diagnostic survey was carried out in early 1973, having been preceded by a limited pilot survey in late 1972. The pilot survey, based on a mail questionnaire, was administered to a small sample of 91 employees at the University of British Columbia. The sample was randomly drawn from the University payroll records and included all categories of salaried employees, of whom only 35 per cent represented teaching staff. The response rate was 68 per cent.

The response to the pilot survey suggested various ways in which the presentation of the questionnaire might be improved, and established that respondents were able to give unbiased answers to the kind of questions being asked. Only one or two of the 62 completed questionnaires had to be discarded because the respondent could not, or would not, understand the 'rules of the game'.

The next step was to administer the improved questionnaire to a larger, and more representative, sample. The British Columbia Provincial voters' roll was chosen as the main sampling frame since an election had been held in the autumn of 1972 and the lists had been recently revised. For the purpose of drawing the sample, and to test whether there was any geographical bias in the responses, the sample was first stratified into three areas: Metropolitan Vancouver, Burnaby, and West and North Vancouver. A fourth group, representing University employees, was later added, because detailed income data were available for part of this group.

The samples within each of the three selected areas were drawn by stratified random sampling. A random number table was first used to select a limited number of polling districts (each containing between 200 and 400 registered voters), after which 30 names were randomly selected from the voting register in each district. The University sample was randomly selected from the University payroll records as before.

The survey was administered to the selected sample by means of a mail questionnaire. The details of the response, which represented an over-all rate of 37·5 per cent, are shown in Table 1.1.

The final samples were made up as follows:

Area	Number of Polling Districts	Size of sample	Size of sample as per cent of	
			selected polling districts	total registered electorate in the area
Metropolitan Vancouver	33	990	8·17	0·43
Burnaby	17	510	9·20	0·62
W. & N. Vancouver	17	510	9·40	0·70
University	–	654	–	–

One of the first things that had to be tested was the possibility of sampling/response bias. There were a significant number of respondents with 'address unknown', and it was also feared that the response rates might be related to some family or socio-economic characteristic. The areas with the highest incidence of 'address unknown' were examined first. Graphs were drawn of the number of unknown addresses in each polling district against selected indicators which might have been related to them. Of these, the percentage of dwelling units rented in each area compared with those which were owner-occupied and detached (this roughly measures the amount of family accommodation in an area) seemed

TABLE 1.1 *Response to mail questionnaire*

	Burnaby	West and North Vancouver	Metropolitan Vancouver	University
Initial sample size	510	510	990	654
Address unknown, etc.*	46	39	118	3
Delivered sample	464	471	872	651
Questionnaire does not apply**	16	13	32	2
Questionnaire rejected***	11	13	24	12
Mode of travel — car	98	103	142	243
— bus	11	18	47	33
— other	10	15	25	54
Direction of travel by car and bus — to work	55	63	87	138
— from work	54	58	102	138
Response	146	162	268	344
Response rate (per cent)	31·5	34·4	30·7	52·9

 * No such address, no such person, moved, building burnt down or demolished
 ** Changed place of work in last four weeks, unemployed, retired, no fixed place of work
 *** Incompletely filled out, travel time too small, incorrectly filled out, answer applies to business travel
  Burnaby: North, Willingdon, Edmonds; West and North Vancouver: West, Capitano, Seymour; Metropolitan Vancouver: Burrard, Point Grey, Little Mountain, South, East

to be the most important. The relationships were by no means exact, but they did show that (a) with only three exceptions, areas with six or more unknown addresses were areas in which more than 50 per cent of the dwelling units were rented and (b) with only four exceptions, areas with six or more unknown addresses were areas in which less than 58 per cent of the dwelling units were owner-occupied and single-detached. A similar graph, plotted with an index of ethnic origin as the independent variable, showed no relationship at all. It therefore seemed reasonable to suppose that the number of unknown addresses reflected no more than the natural turnover of population and did not represent an explicit form of bias.

The other test for bias was applied to respondents. It was performed by comparing the response rates in each area to indices of family status and to various socio-economic characteristics (8). The former included statistics relating to family structure, type of education, and housing characteristics. The latter included occupational information, information on household expenditure, and ethnic origin. Graphs were again plotted of the response rate in each area against each of these two indices, and again they showed no evident bias.

Finally, before any detailed analysis took place, the completed questionnaires were edited, but with care to ensure that no record was excluded on the grounds that the respondent simply appeared to be 'irrational'. This was a very real danger since the criteria of rationality often consist of excluding those respondents who do not conform to the researcher's own expectations about behaviour. To avoid this potential source of bias, the amount of editing was kept to a minimum and consisted only of:

(a) removing records which applied to business travel or which were incorrectly, or ambiguously, filled out;

(b) removing records where the over-all length of journey was too small (e.g. less than a ten-minute bus ride) or where the record contained too few entries to make processing worth while.

This eliminated 7·3 per cent of the car records and 11·5 per cent of the bus records. The numbers of accepted car and bus records were thus 584 and 109, and by visual inspection it was judged that 84 per cent of the former and 90 per cent of the latter were fully filled out (the remainder had responses to about half the questions).

1.3.3. QUESTIONNAIRE DESIGN

The questionnaire was presented to respondents in three sections. There were two questionnaires,—one pink and one blue, and a covering letter. The letter gave a brief description of the survey, followed by the question 'Have you changed your place of work during the past 4 weeks?' This was designed to weed out atypical respondents. If their answer was affirmative they simply returned the covering letter. The letter then directed car users to the pink questionnaire and the bus users to the blue one; 'other modes' (which included car and bus, car passenger, bicycle, walk, hitchhike, etc.) were provided for on the covering letter itself. The coverling letter also gave the definition of 'normally' as 'more than half the time'.

The design of the two questionnaires was intended to satisfy the following requirements:

(a) The questions should relate to decisions with which the respondent was expected to be familiar. In the bus questionnaire this was achieved by relating savings in time to the possible use of dial-a-bus and express buses, both familiar concepts to bus users in Vancouver. The questions in the car questionnaire were a little more unreal. The reductions in walking times were related to 'premium' parking lots, and this was quite familiar and realistic. The question of in-vehicle speed, however, was dealt with as follows:

If, by means of synchronized traffic lights and by limiting access from side streets, some existing streets were turned into express roads with a speed limit of 50 mph, what toll would you consider it just worth paying . . . etc.

The question was intentionally not phrased in terms of an urban motorway because of the general antipathy towards such roads and the very large difference in quality (accident hazard, etc.) between motorways and the current urban arterials in Vancouver. But it still did not exclude the possibility of bias. The type of change envisaged, i.e. an urban clearway with a higher speed limit is fairly common, but it does result in more noise, community severance, etc. and, more importantly, it tends to confront the motorist with a *fait accompli*. If all principal arterials were dealt with like this, the motorist would have to use them and pay, whether he wanted to or not. This could clearly affect his response to the question, since he might interpret the suggestion as one that would diminish his range of travel choice.

(b) Use should be made of the familiar units in which people are accustomed to record times and costs. Travel times were therefore always asked for in terms of 'does it take less than 5 mins, 5 to 10 min, more than 10 min, etc.', while costs were always presented in the equally familiar terms of, '10c, 25c, 50c, 75c, $1·00, etc.' Now the choice of these units did create some problems, because it is usually thought that the scale chosen to represent the variables on a questionnaire is likely to bias the response. A well-designed sample will therefore usually choose several scales and will randomize them among the questionnaires, but unfortunately this was not possible in the present case because of the limited range of the scale. Some inferential evidence from the pilot survey, however, suggested that the bias was relatively unimportant. The same scale had been used for both the bus and the car questionnaires, and yet the preference functions for each turned out to have completely different shapes. A significant scale bias should have led to more similarities.

(c) A system of coding should be provided so that individual respondents could, if necessary, be identified for further analysis. This was done simply by putting code numbers on each questionnaire and matching these with the individual's polling number, though in surveys of a more confidential nature, this procedure would clearly have been unacceptable.

Half the questionnaires sent out dealt with the journey *to* work; the other half with the journey *from* work. An equal number of each type was administered to each polling district.

1.4. RESULTS OF THE SURVEY

The results of the survey are presented in three parts corresponding to the objectives set out in 1.3.1.

1.4.1. RESPONDENT BIAS IN STATED TRAVEL TIME CHARACTERISTICS

The result of the travel time survey is summarized in Fig. 1.4. It shows, for each of the five segments comprising the work journey, the actual time reported by bus users and the estimated time by bus reported by car users whose 'other regular means of transport' was using a bus (that is, 91·5 per cent of all car users).

The significant and consistent feature of this Figure is that, without a single exception, the average bus times reported by car users are substantially higher than those reported by bus users. Some of these differences might be explicable. For example, the 'car user' category might include a large number of people who live or work in areas poorly served by public transport. Indeed, it may well explain *why* they belong to this category, leaving only those people who are well served by public transport as potential bus users.

This might be a reasonable explanation. The difference cannot simply be explained by a lack of knowledge. If car users simply know less about bus travel, one would expect to find some error in their responses, but only a random error without this substantial and consistent bias. Only in the case of 'waiting time' might the bias be explained in terms of uncertainty. Since car users rarely, if ever, use buses they will probably know less about scheduled timetables and will therefore be more prone to waiting 'errors' than regular bus users. Regular bus users will plan to reach the bus stop shortly before the advertised time of departure:

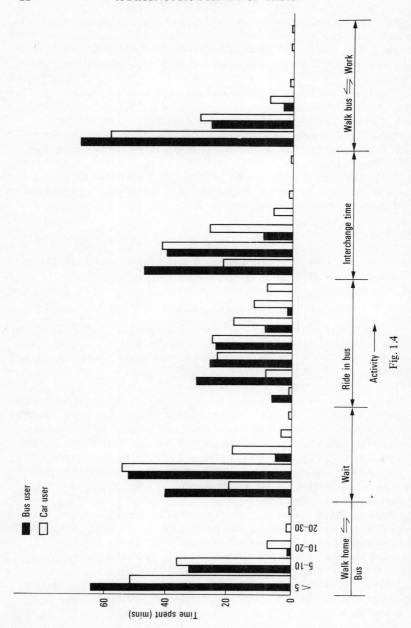

Fig. 1.4

infrequent users will probably arrive at random and simply wait until the next bus arrives. The remaining bias, particularly that relating to walking times and interchange times, can only be explained by showing that car users and bus users generally have a different type of service available to them.

The above hypothesis was tested by matching bus and car records from the same sampling zones. The matched records apply to journeys with common origins, but do not necessarily apply to common destinations. Eight of these matchings are shown in Table 1.2. The records selected are more exaggerated than most, but are still representative of the kind of bias present in most records. The zones from which the bus and car users come are clearly fairly small. It is possible to walk across most zones in about 5 min and from end to end of them in about 10 or 12 min. And yet, although all but one of the bus users shown in these records reported that it took them less than 5 min to walk to the bus stop, the car users gave answers varying from less than 5 min to as much as 30 to 40 min! The mean of their estimated walking times was 12·5 min compared with a mean of just over 2·5 min for the bus users. There is no easy way of explaining these consistent differences. The small size of the zones and the response from bus users in these zones clearly place the proximity of the bus stops within the range of the car users' spatial experience. The pattern of bus services (which are generally organized on a rectangular grid basis) likewise means that the effect of differences in destination—which might otherwise have implied catching buses operating different routes—is minimal. One possible explanation for these discrepancies, which show the same pattern of bias throughout, is that of ex post rationalization. The bus walking times and interchange times should, with few exceptions, be similar for both car and bus users in each zone. The responses suggest, however, that car users give a biased response to justify their reasons for using a car, although the effect might also be attributable to a perceptual bias linked to the poor image of bus services held by the car users.

The finding, together with Quarmby's (2) result that there was remarkable agreement between car users and bus users on the speed and time of car travel, suggests that empirical studies of modal choice based on stated information about preferred and alternative modes lead to a significantly biased result. Indeed, the suggested process of ex post rationalization brings into question the whole concept of explaining modal choices in terms of relative costs and times. These factors are probably important, but relative time clearly does not enter into the individual's decision process in the simple way that current modal choice models invariably assume.

TABLE 1.2 *Bus travel times recorded by bus users and estimated by car users*

Zone	User	Travel Time (min)						Size of zone (miles)
		Walk, bus to/from home	Wait for bus	Walk, bus to/from work	Wait for connecting bus	Ride in bus	Over-all car time for journey	
01:17	Bus	<5	10−20	<5	<5	20−30		0·2 × 0·5
	Car	<5	20−30	30−40	30−40	>60	25	
02:70	Bus	<5	<5	5−10	5−10	30−40		0·2 × 0·7
	Car	30−40	20−30	<5	10−20	>60	35	
	Car	5−10	20−30	5−10	20−30	>60	35	
03:72	Bus	5−10	5−10	<5	−	20−30		0·13 × 0·5
	Car	<5	20−30	<5	−	>60	30	
03:79	Bus	<5	5−10	5−10	<5	30−40		0·75 × 1·0
	Car	10−20	40−60	>60	>60	>60	40	
06:44	Bus	<5	<5	5−10	−	5−10		0·3 × 0·13
	Car	<5	20−30	<5	20−30	40−60	25	
09:115	Bus	<5	<5	5−10	<5	20−30		0·25 × 0·5
	Car	10−20	20−30	5−10	−	30−40	25	
10:65	Bus	<5	5−10	<5	<5	10−20		0·2 × 0·3
	Car	5−10	20−30	5−10	10−20	>60	45	
10:113	Bus	<5	5−10	5−10	<5	20−30		0·3 × 0·3
	Car	10−20	10−20	20−30	10−20	40−60	35	

− = no changes made

1.4.2. BIAS IN STATED TRAVEL COSTS

It is not easy to frame a question about travel costs that will lead to an unbiased response. Questions like 'How much does it cost you to use your car?' usually produce a response (e.g. 78·5 per cent of the respondents answered the questions on car costs), but the figures given do not necessarily represent the figures present in respondents' minds when modal choices are decided. In most cases the stated figure is simply a construct, derived only *after* the question has been asked.

This creates a serious problem for survey methodology. How does one get behind the individual's normal response mechanism, which provides an automatic positive response to certain types of question, to discover what his 'real' decision parameters are? The questions on car costs included in the present survey clearly did not attempt to answer this question definitively, nor to try and probe which cost parameters really did enter into the individual's decision process. The objective was more limited. The questionnaire simply tried to find out (a) whether, and in what way, people responded to the questions on car costs and (b) whether their responses provided a reliable basis for calculating the 'perceived cost' of using a car. This was done by asking 'How much do you estimate it costs you to use your automobile to travel to and from work?' The question was further subdivided into: parking charge, cost of using car, contribution from passengers (if any), and other expenses.

Inspection of the completed questionnaires revealed four basic types of response. First, came the non-numerical responses (nearly 22 per cent of the total) which were fortunately not entirely blank. They contained no numerical response (or one too vague for coding), but were rich in unsolicited comments like 'don't know', 'impossible to estimate but minimal', 'not much', 'not certain, possibly . . .', '?', and so on.

Second, came the formal responses in which the individual took an accepted figure for car costs per mile (possibly because he was sometimes paid a mileage allowance) and multiplied it by the daily distance travelled. The most common figures appearing on the questionnaires were 6c, 10c, and 12c per mile.

Third, came the calculated responses. These were given only by a minority, but presented (presumably for the respondents' own use) elaborate calculations of cost of fuel divided by m.p.g. multiplied by mileage travelled, annual maintenance expenditure divided by average annual mileage multiplied by work journey mileage, and in one case even the average cost of traffic fines per mile travelled. This was then used to calculate a cost for the work journey.

Finally, came the erratic responses. The respondent simply seemed to pull a figure out of the air—usually rounded to a whole number of dollars—and wrote it down.

The variability of these responses is further confirmed by Table 1.3, which compares estimated car costs with in-vehicle times. It shows a very wide dispersion, with a significant number of people either grossly underestimating car costs (9 per cent of the completed records state a car cost less than fuel costs), or grossly overestimating them (18 per cent of the completed records state a car cost in excess of the cost of fuel, maintenance, and depreciation). Some of the higher figures might be explicable in terms of the respondent's method of calculating his costs. The costs that are lower than fuel costs, on the other hand, can only really be understood if they are interpreted in terms of a decision model that does not have to rely on the motorist being consciously aware of the relative costs of competing transport modes. Indeed, the figures in Table 1.3 tend to underestimate the size of the group stating car costs less than fuel costs. The figure of 2·5c per mile on which the Table is based relates to a U.S. compact car in urban conditions. The average car used in Vancouver probably consumes

TABLE 1.3　Comparison between estimated car costs and travel time

Estimated cost of using a car to go to and from work (cents)

Length * of journey (min)	<10	10–20	21–40	41–60	61–80	81–100	101–150	151–200	>200	Grand Total
Less than 10	2	6	13	16	6	5	6	2	3	
10–20	9		26	59	34	40	12	10	10	
20–30	1		4	22	11	25	14	20	9	
30–40	–		2	14	7	13	8	12	8	
40–60	–		1	3	2	6	5	4	7	
More than 60	–		–	1	–	–	–	–	–	
Costs										
Less than fuel cost **	12		7	18	2	–	–	–	–	39
higher than normal running **	–	–	–	–	6	5	18	32	22	83
Costs reasonable	6		39	97	52	84	27	16	15	336
Total	18		46	115	60	89	45	48	37	458

* In-vehicle time each way

** Fuel cost taken as 2·5c per mile; normal running cost as 10c per mile (excluding driver's time)

between 3c and 4c of fuel per mile. The figure of 9 per cent grossly underestimating car costs is therefore conservative.

The average perceived cost works out at 7·7c per mile. This is less than the total cost of running a car in Vancouver (approximately 10–12c per mile), but is fairly close to its marginal cost (fuel, tyres, routine maintenance, wear and tear, and lubricants). Modal choice inquiries which ask respondents to state how much it costs them to run their car therefore often conclude, on the basis of a mean value which is roughly equal to the marginal cost of operating a car, that relative costs constitute an important decision variable and are fundamental to any explanation of modal choice. The evidence above casts doubt on this assumption. All the respondents realize that it costs them something to run a car, but nearly all seem to respond with the spirit of the answer 'not certain, possibly . . .' The evidence suggests that stated car costs do not give a true indication of the monetary content of any decision parameters associated with car usage and that conclusions based on 'perceived cost', as conventionally calculated and used, are probably fallacious.

It is also quite clear that the consistent bias in stated travel times, and the widespread ignorance of car costs, challenge the very fundamentals of current theory. People simply do not seem to choose between alternative modes of transport on the basis of an explicit trade-off between the relative times and costs of each mode.

There are really two areas of doubt. First, it seems that the usual methods of estimating modal choice models (and route choice models for that matter) do not provide an accurate or unbiased measure of the imputed value of travel time savings. Second, the apparent existence of ex post rationalization for modal choice and the secondary role of car running costs as a decision parameter imply that modal choices are based on a behavioural process different from the one postulated. Relative times and costs *might* be important in certain circumstances, or to certain individuals, but they clearly do not provide a complete explanation of modal choice.

1.5. VALUE OF TRAVEL TIME SAVINGS

The two sources of bias examined in sections 1.4.1. and 1.4.2. suggest that the implied values of travel time savings derived from conventional modal and route choice models are incorrect. Quite apart from the problem of group averaging implied by the estimation procedure, these models will generally overestimate time values because the bias in car costs seems to be much greater than that in relative travel times.

The present study avoids this bias by concentrating on behavioural responses within a single mode. By doing so it runs the risk of introducing other sources of bias, but given the diagnostic nature of the inquiry, any such bias should be relatively unimportant: though it might distort the precise values of some parameters, it should not affect the general functional relationships.

1.5.1. EFFECT OF SIZE OF SAVING AND JOURNEY LENGTH

The first task of the inquiry was to examine the relationship between the value of a time saving, V, the size of the time saving, Δt, and, where possible, the over-all length of journey T. The relationship between the mean values of V and each of these parameters was plotted graphically, for both bus and car users, using in-vehicle times as one of the independent variables. It was not possible to do the same for walking and waiting times because of the limited number of observations exceeding 10 min. Indeed, walking to and from a car park in the case of car users did not provide any meaningful graphs at all. Over 92 per cent of car users stated that their walking times were less than 5 min. Graphs for walking and waiting times were therefore plotted of V versus Δt for bus users only.

The graphs for car and bus users are shown in Figs. 1.5, 1.6, 1.7, and 1.8. Fig. 1.5 shows the relationship between V, Δt, and T for in-vehicle times. In-vehicle time was chosen as an independent variable, in preference to over-all journey time, because the statistical correlations between V and the two different measures of journey length suggested that in-vehicle time was more significant. The statistical (Kendall) correlations were as follows:

		Correlation	*Significance*
Bus users:	V versus in-vehicle time	0·104	(0·009)
	V versus computed over-all journey length	0·086	(0·026
Car users:	V versus in-vehicle time	0·196	(0·001)
	V versus computed over-all journey length	0·189	(0·001)
	V versus stated over-all journey length	0·190	(0·001)

Each line in Fig. 1.5 shows the relationship between V and Δt for a particular value of T. The dotted line shows the simple relationship between V and Δt. Fig. 1.6 shows more clearly the relationship between V, Δt, and T, while Fig. 1.7 tests the hypothesis that V is related to $\Delta t/T$, i.e. the proportion of travel time saved. Fig. 1.8 shows the relationship between V and Δt for walking and waiting times.

The figures show a number of interesting relationships:

(a) V is generally a non-linear function of Δt. Indeed, when the effect of journey length is ignored, it seems to be a function of Δt^2.

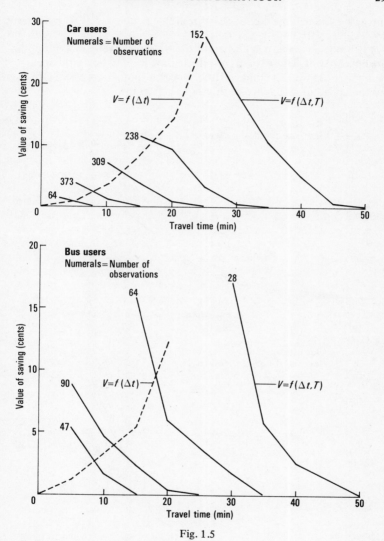

Fig. 1.5

(b) Both bus and car users exhibit some form of threshold behaviour. It is most marked in the case of car in-vehicle time which exhibits a discrete threshold giving rise to a discontinuity in the time-cost preference function. The bus users also exhibit threshold behaviour, but of a more continuous nature.

(c) For car users, the nature of the threshold changes as journey length increases. Indeed, the size of the threshold (5 min at $T = 50$ min) seems to decline continuously as T falls.

(d) It is not possible to say whether the time-cost preference functions are linear or not. Those for car in-vehicle time seem to be nearly linear: those for bus in-vehicle time might be linear, although the effect is masked by the apparent difference in threshold behaviour.

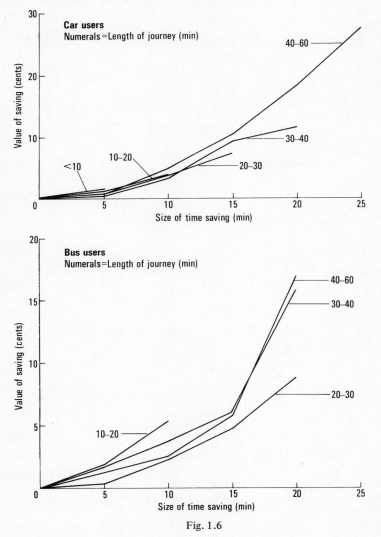

Fig. 1.6

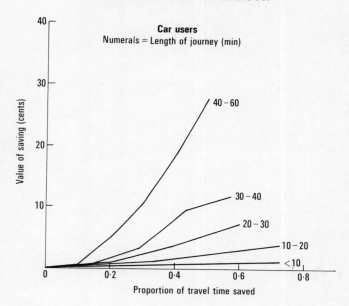

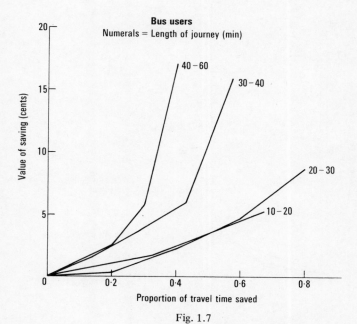

Fig. 1.7

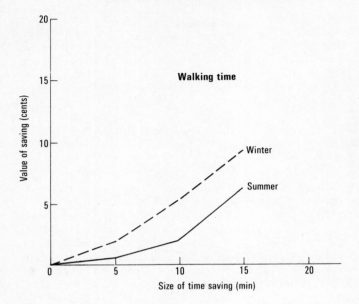

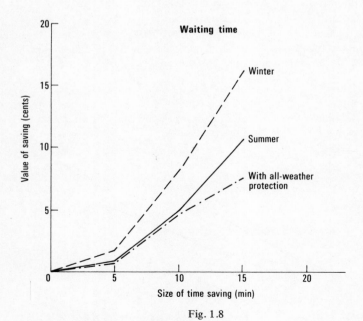

Fig. 1.8

(e) Journey length (in-vehicle time) seems to be significant. Except for short journeys, the marginal value of time savings increases with journey length. This implies that the perceived quality of the journey declines, or that the opportunity cost of the journey time rises, as its over-all length increases.

(f) V for in-vehicle time does not seem to be a simple function of $\Delta t/T$.

(g) When the effect of journey length is ignored, the functions $V = 0.03\Delta t^2$ and $V = 0.036\Delta t^2$ seem to give a reasonable fit to the two sets of bus and car in-vehicle data. This implies that the marginal value of in-vehicle time is more or less the same for each group. However, if V is treated as a linear function of Δt, and the range of observations is confined to 15 min or less, the marginal value of in-vehicle time for bus users appears to be only about half of that for car users (compare Lee and Dalvi's (5) ratio of approximately 3).

(h) The graphs of walking and waiting times conform to the same general shape as those of V versus Δt for in-vehicle time.

(i) The marginal value of savings in walking and waiting times in summer is lower than in winter (in Vancouver this corresponds roughly to dry versus wet weather). The apparent effect of all-weather protection at bus stops is to cause users to revert to a behaviour pattern that exhibits marginally less disutility than the pattern for summer behaviour.

(j) If the differences between winter and summer weather are ignored, the relationships between V and Δt for walking and waiting times can be approximated by $V = 0.04\Delta t^2$ (walking) and $V = 0.06\Delta t^2$ (waiting). This implies that if journey length is ignored the marginal values for walking, waiting, and in-vehicle times for bus users are in the proportions 1.33 : 2.0 : 1.0.

The above results not only confirm a number of our intuitive notions about the way people might be expected to behave under these conditions, but clearly have widespread implications for the way in which time savings are used in any economic evaluation.

1.5.2. EFFECT OF DIRECTION OF TRAVEL

The next variable tested was the direction of travel. Results for both bus and car users were classified according to whether the journey was being made *to* work or *from* work. The results are plotted in Figs. 1.9 and 1.10.

The results for bus in-vehicle time are unfortunately rather poor, because of the limited size of the sample. This is particularly true of the longer journey lengths, where the curves are rather meaningless. On the shorter journeys, where the sample size is big enough, there does not seem to be any marked difference between marginal values and the direction of travel.

The results for car in-vehicle time are more informative (and reliable)

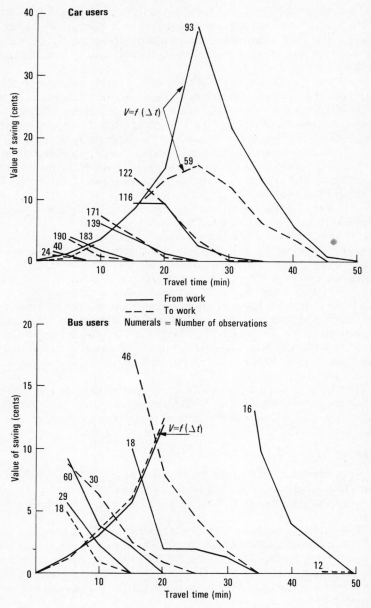

Fig. 1.9

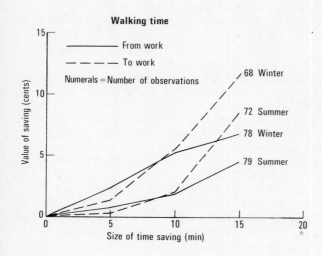

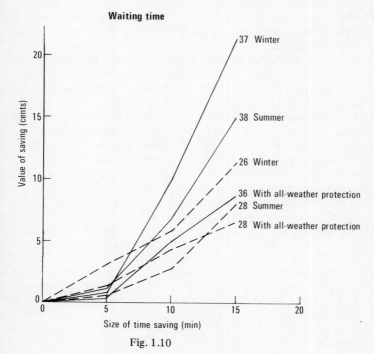

Fig. 1.10

because of the very much larger samples involved. On the shorter journeys, i.e. those less than 40 min, there seems to be no relationship between marginal values and the direction of travel. On the longer journeys, on the other hand, i.e. those with journey lengths of 40–60 min, there does seem to be a significant difference. Time saved going home seems to be worth roughly twice as much as time saved going to work! This is not too surprising, since the opportunity cost of time is probably higher at the end of the working day, because of the wider range of substitute activities available. However, since a significant difference only appears on the longer journeys the observed effect may very well be due to fatigue. This effect clearly requires further investigation, particularly in relation to other journey lengths.

Fig. 1.10 shows the effect of trip direction on the marginal value of walking and waiting times. These results are even more surprising. The effect of trip direction on walking time is exactly the opposite to its effect on waiting time (which is consistent with the observed effect on car in-vehicle time). In the case of walking time, the marginal value of time savings going *to* work exceeds that coming *from* work; in the case of waiting time the marginal value of time savings coming *from* work exceeds that going *to* work! The differences are furthermore quite marked (in the case of waiting time they differ by a factor of nearly 2) and maintain the established relationship between the summer, winter, and all-weather values.

The only noticeable difference is that the effect of trip direction on waiting time is more marked than it is on walking time and this suggests that there might be a secondary effect operating on the marginal value of walking time. The higher marginal values for waiting time coming from work can probably be explained in terms of time-of-day. Waiting time from work will generally be made in the evening when bus stops are less secure and congenial than they are in the morning. The opportunity cost of time and the effect of fatigue—following the arguments put forward for car in-vehicle time—is also likely to be higher at the end of the working day. What then reverses this effect on walking time?

One possibility is that walking times do exhibit this effect, but that it is masked by a more powerful secondary effect. Urban life in North America, and Vancouver is no exception, tends to be rather sedentary, providing few opportunities for casual exercise. People walk far less than they do in Europe (e.g. over 92 per cent of car users and 66 per cent of bus users reported walking times to and from car parks or bus stops of less than 5 min) and, partly as a result of this, tend to be more overweight. It is therefore conceivable that the short walk on the homeward journey after a sedentary day's work might be viewed as a substitute for a brief spell of exercise. This would not apply in the morning when the desire for exercise would probably be less.

However, this explanation, though plausible, is still only conjectural. All that can be asserted with confidence is that trip direction does seem to affect walking and waiting times and that the effect on the former seems to be quite different from the effect on the latter.

1.5.3. EFFECT OF CONGESTION

The next variable tested was the effect of traffic congestion, the test being applied only to car in-vehicle time. The four levels of congestion specified on the questionnaires were combined into two groups: free flow plus little congestion, and mild congestion plus severe congestion. Separate time-cost preference functions were then plotted for each of these groups. The results are shown in Fig. 1.11.

The effect of congestion is not very marked, not large enough to warrant estimating time values as an explicit function of congestion. But it is possible that the small observed differences in Vancouver are attributable to the generally low level of congestion experienced by car users. Congestion does occur, and is particularly acute on those routes with bridges or tunnels, but it is nevertheless fairly localized and only really affects travellers moving between (rather than within) the major municipalities. Indeed, in the present sample, only the journeys from North and West Vancouver to Metropolitan Vancouver and Burnaby experience real and persistent congestion. Within Metropolitan Vancouver and Burnaby, or between them, it is quite mild and localized. The curves of Fig. 1.11 therefore probably do not give a very accurate measure of the effects of congestion. They do indeed show that it

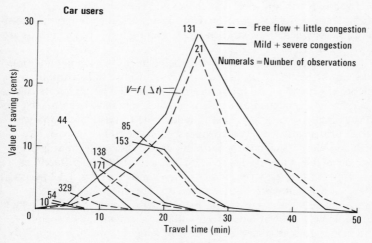

Fig. 1.11

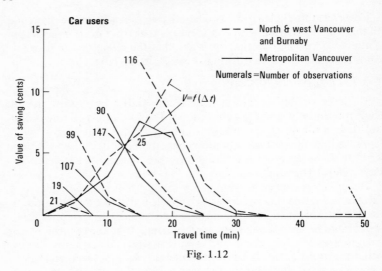

Fig. 1.12

affects marginal time values and that the effect increases with the level of perceived congestion, but the observed differences are probably less than one would find in a major urban centre where congestion was more widespread.

1.5.4. EFFECTS OF RESIDENTIAL AREA

This effect was tested by classifying car users into two groups: those who lived in North and West Vancouver and Burnaby; and those who lived in Metropolitan Vancouver. Time-cost preference functions, classified by journey length, were then plotted for each group. The results are shown in Fig. 1.12.

It was thought that the area of residence might affect marginal time values on the grounds that people who chose to live in Metropolitan Vancouver generally paid more for housing but had better access to a wide range of civic amenities, and might therefore have been paying (through higher property prices) to reduce the travel time component of their urban activities, while people who chose to live in areas like North and West Vancouver or Burnaby were consciously trading off lower property values against reduced accessibility to civic amenities. The two sets of time-cost preference curves shown in Fig. 1.12 nevertheless show a quite different relationship. The difference between them is not very great, but the nature of the difference does imply that the people with a lower level of accessibility seem willing to spend *more* money to save time than those with a higher level of accessibility! Clearly, if this conclusion was validated by more rigorous experiment, it would have important implications for many models of urban growth.

1.5.5. EFFECT OF INCOME

This was one of the most important variables tested in the survey. Detailed gross personal income data were available for 76 car users in the University sample. They represented a wide range of professions and income groups, as shown in Table 1.5. Only 41 per cent were classified as academic staff, the remainder being secretaries, technicians, maintenance workers, and clerical staff. The only evident bias was that their mean income was slightly higher than average.

The 76 respondents were randomly selected from the University sample of 210 accepted car records, i.e. 36 per cent of the University car records. To test how representative this sub-sample was, time-cost preference curves were plotted for the whole university (210 respondents) and for all the other respondents. These are shown in Fig. 1.13.

The preference curves in this graph do show some bias, relatively unimportant (and in a reverse direction) on the shorter journeys, but significant on journeys over 30 min, which may be due to the high incidence of white-collar workers in the University sample. This will not necessarily invalidate any conclusions about the effect of income on

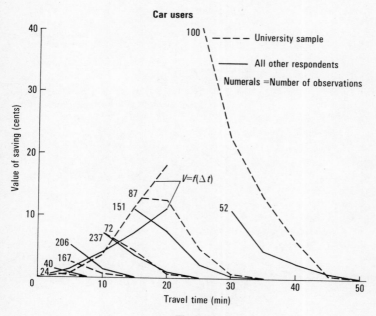

Fig. 1.13

marginal time values: it will simply lead to a biased estimate of any
general relationship between marginal values and wage rates.

The effect of income on marginal values was examined in the
following way. Graphs were plotted of V versus Δt for different levels
of income. Because of the limited number of records available, only
two income groups were distinguished: those people whose income was
$900 or less; and those whose income was more than $900. However,
once the two curves had been plotted it was apparent that the difference
between the two groups might be attributable to the differing incidence
of long and short journeys. So separate graphs were drawn showing the
values of the two income groups (a) for all journeys and (b) for journeys
longer than 30 min. The results are set out in Fig. 1.14.

The effect of gross income levels seems to be quite opposite to that
assumed in all other empirical studies. The *highest* marginal values are
displayed by the *lowest* income group! The separation of the sample
into a further sub-group, i.e. those people making journeys longer than
30 min, likewise confirms the effect of journey length on marginal values
and shows that the observed income effect is independent of journey
length. Yet this result was not entirely unexpected. The presumption
of economic theory that income and marginal values are positively

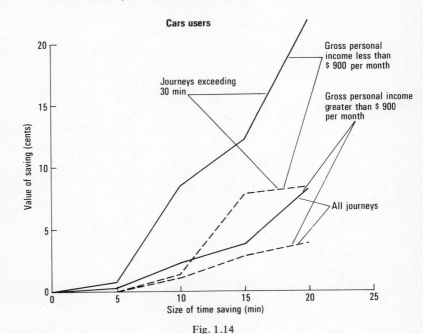

Fig. 1.14

correlated is based on the concepts of richness and poorness: a 'rich' person is presumed to be willing (and able) to spend more than a 'poor' person to save travel time. What Fig. 1.14 shows is that gross income is an extremely poor index of richness or poorness. Indeed gross income and marginal values seem, if anything, to be negatively correlated, although this is due to fortuitous circumstances rather than to any fundamental causal relation. The basic relationship is between disposable income and marginal values. The low-income members of this sample seem to be young, are either unmarried, or married without children and with a working spouse, and living in rented accommodation. They therefore have a relatively high disposable family income in relation to their household commitments. The high-income members, on the other hand, tend to be male family heads (87 per cent males in this group; 49 per cent in the lower-income group) with non-working wives and very substantial family commitments (mortgages, credit cards, etc.).

TABLE 1.5 *Professional status and income group of selected University respondents*

Status	No.	Income level *($ p.m.)*	No.
Academic staff			
All Professors	27	<500	12
Lecturers & Instructors	2	500–750	17
Misc. & Administrative	2	750–1000	21
		1000–1250	8
Non-academic staff (income level ($ p.m.))		>1250	18
<500	12		
500–750	16		
>750	17		
	76		76

The notion that marginal time values are positively correlated with income need not therefore be totally rejected; what must be rejected is that they are related to gross personal income. The common practice of expressing time values as a fixed proportion of the wage rate is therefore incorrect. This conclusion is supported by other evidence. The Local Government Operational Research Unit (LGORU) in two recent studies concluded:

Dr. Quarmby shows that, for the Leeds data, the value of time as a proportion of income was reasonably constant across income groups. Our work on this on the pooled data was inconclusive, possibly because our samples were not large enough, when divided into subsets to allow for the high variability across towns. (9)

and

The model hypothesising that the value of time is proportional to income gave a calibration at an acceptable level of significance, but the model is not as good as the 'traditional' model which implies a constant value of time. It is not therefore possible to regard this relationship as meaningful. (3)

It is therefore quite clear that the marginal value of travel time savings is not a fixed proportion of the wage rate and that a better index of disposable income, e.g. family structure or stage in the life cycle, must be developed before separate group values can be validly used in applied planning studies.

1.6. STATISTICAL ANALYSIS

The information collected in this survey, since it is not continuous, does not lend itself to rigorous statistical analysis, and the diagnostic nature of the inquiry makes the use of unduly complicated methods of analysis (e.g. multiple regression) inappropriate. In spite of these reservations it was nevertheless decided to compute the Kendall correlation coefficient between the marginal value of time savings and each independent variable. The results of these calculations are shown in Table 1.6. The Kendall, rather than the usual Spearman coefficient, was used because of the discrete nature of the data.

Table 1.6 confirms the results of the more informal analysis outlined above. In the case of in-vehicle times, the size of the saving (Δt) and the over-all length of in-vehicle time (T) are both highly significant. The level of congestion for car users is also highly significant. Residential area likewise seems to be significant for bus users. However, since the quality of bus services varies quite significantly between areas (and is more homogenous within areas), this is of no great importance.

In the case of walking and waiting times, the size of the saving is again highly significant. The over-all length of walking time seems to be significant in winter; while that for waiting time seems only to be moderately significant in summer (or with all-weather protection). Area of residence is even less significant, although it does show some significance in the case of waiting in summer (or with all-weather protection), while direction of travel only seems significant for waiting with all-weather protection.

The statistical analysis therefore confirms the informal analysis set out above. It simply reinforces the importance of the size of saving, over-all length of in-vehicle time and level of congestion experienced, but suggests that direction of travel and residential area may not be as important as the informal analysis suggests.

TABLE 1.6 Kendall rank order correlation coefficient between marginal time values and tabulated variables

Activity	Size of saving	Length of activity	Area of residence	Direction of travel	Computed over-all journey length	Stated over-all journey length	Level of congestion
Car in-vehicle time: N = 1333	0·348 (0·001)	0·196 (0·001)	-0·0114 (0·267)	0·006 (0·375)	0·189 (0·001)	0·190 (0·001)	0·203 (0·001)
Bus in-vehicle time: N = 230	0·337 (0·001)	0·104 (0·009)	-0·092 (0·019)	0·038 (0·195)	0·086 (0·026)	—	—
Walking time: summer N = 129	0·134 (0·012)	-0·024 (0·344)	0·020 (0·370)	0·044 (0·228)	0·031 (0·300)	—	—
Walking time: winter N = 127	0·278 (0·001)	0·163 (0·003)	0·040 (0·252)	0·010 (0·433)	-0·055 (0·181)	—	—
Waiting time: summer N = 30	0·444 (0·001)	0·200 (0·060)	0·196 (0·064)	-0·033 (0·398)	0·208 (0·053)	—	—
Waiting time: winter N = 26	0·311 (0·013)	0·085 (0·271)	-0·020 (0·445)	0·183 (0·095)	0·180 (0·098)	—	—
Waiting time: with all-weather protection N = 29	0·529 (0·001)	0·290 (0·064)	0·183 (0·082)	0·228 (0·041)	0·180 (0·086)	—	—

N = number of observations
Figure in brackets is level of significance

1.7. IMPLICATIONS FOR THEORY

The most important implication of the above results is that models of modal choice can no longer be credibly characterized in terms of continuous, linear functions of time and cost with constant marginal values.

Another important implication concerns the use of appropriate time values in applied planning studies. The present survey clearly shows that the value of a saving in journey time (on the work journey) is a function of a number of variables which include (a) the over-all size of the saving, (b) the time taken on each stage of the journey, (c) the incidence of general weather conditions when walking and waiting, (d) the direction of travel (in some instances), and (e) the level of perceived congestion. The respondents also seem to exhibit threshold behaviour and show no consistent relationship between marginal values and gross income.

Nearly all these effects can be explained in terms of time-cost preference functions. The functions might very well be linear but are almost certainly not parallel. In the case of work journeys they seem to increase in steepness, i.e. the marginal value of any time saving increases as the amount of time spent in this activity increases.

The slope of a time-cost preference function, for a given duration, also varies with the part of the journey involved. It seems to be steepest for waiting time, followed by walking time, and then by in-vehicle time. The incidence of inclement weather conditions, direction of travel, level of perceived congestion, and the disposable income of the respondent also seems to affect the slope of the function.

If V is a function of T as well as of Δt, the effect of omitting T as a variable gives rise to an apparently non-linear relationship between V and Δt. Indeed, it gives rise to a function like that shown dotted in Fig. 1.15. It is thus no surprise that Thomas and Thompson (4) found a non-linear relationship.

The threshold behaviour is more difficult to explain. It implies that the usual indifference curves of economics change shape in the region of a point of equilibrium. This implication might nevertheless be due to the way in which the theory of indifference is articulated. It usually shows people responding continuously and instantaneously to changes in relative prices subject to precise budget constraints.

The real world does not work in this way. Constraints are fairly imprecise (people sometimes spend more or less money or time than they intended to or 'could afford'), while the general inertia in behaviour patterns—particularly in the short run—suggests that people do not respond continuously to relative price changes. It might be more appropriate to think of indifference curves as defining a propensity to behave in a given way and to interpret this propensity in terms of a band of indifference defining the upper and lower limits of individual behaviour.

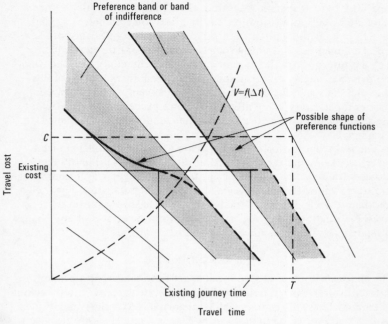

Fig. 1.15

This concept (applied to time-cost preference functions) is also illustrated in Fig. 1.15. It shows how, given certain assumptions about the
effect of increasing journey times (shown dotted), threshold behaviour
could be treated as a movement within a time-cost preference band from
an existing point of equilibrium. However, as the long-term point of
equilibrium moves, so also will the indifference band.

The effect of budget constraints, for both time and cost, can be
illustrated in the following way. Each individual is subject to at least
two constraints, the maximum amount of money and the maximum
amount of time that he can afford to spend on the journey. The two
constraints thus define a feasible area of choice in his two-dimensional
choice plane. The lines C and T, for example, might define such a
feasible area for an individual and would show that certain time-cost
choice combinations, although lying within his general time-cost
preference band, will nevertheless be excluded from his displayed
choices by one or other of the constraints.

A full explanation of urban journey-to-work behaviour clearly requires
extension to other dimensions. Additional modal attributes such as car
availability (not car ownership), parking availability, privacy, and

flexibility require specific attention; as do the ways in which some of these are perceived. It is quite clear that decision (or choice) models have to pay far more attention to the nature and significance of preceived qualities and of the subjective framework through which they are interpreted. The present survey has again confirmed the wide gap between 'vague' car costs and the 'declared' market price of bus fares and daily parking charges. The two cost centres are perceived differently and very likely affect the choice process in quite different ways. Both should therefore be examined, and probably modelled, separately.

The full specification of the behavioural basis for modal choice is nevertheless beyond the scope of this inquiry. It was primarily concerned with only two dimensions of travel: perceived travel time savings and stated values (i.e. how much are these savings worth?). The model can therefore yield some of the parameters needed for purposes of project evaluation, but gives little insight into the effects of any change in relative travel times and costs on modal choice.

1.8. CONCLUSION

The above survey has not produced exact time values appropriate for different travel time situations. That was not its purpose. It was seeking an empirical basis for a more general theory of travel time valuation to replace that indirectly imputed from observed modal choice behaviour. On that count it has largely succeeded. It has shown that modal choice decisions are based on a far more complicated perceptual and behavioural process than has hitherto been assumed and that the method of estimating current models (principally via discriminant analysis) produces biased results.

The method of explaining time values in terms of time-cost preference functions seems quite capable of illustrating the major empirical findings of this survey. In terms of importance these findings are:
threshold behaviour;
the weakness of gross income as an index of richness and poorness;
the effect of weather conditions on the values of walking and waiting times;
the probable importance of the length of time spent in each activity; and
the possible importance of journey direction.
The specific effects of these factors have not been accurately tabulated. The diagnostic survey technique has nevertheless shown that each seems to have an important—and behaviourally consistent—effect and, if it has done nothing else, it has at least indicated in which direction future travel time research should move.

REFERENCES

1 Beesley, M. E., The Value of Time Spent in Travelling: Some New Evidence, *Economica*, 32, pp. 174–85, 1965.
2 Quarmby, D. A., Choice of Travel Mode for the Journey to Work: Some Findings, *J. Transpt. Econ. Policy*, 1 (3), pp. 273–314, 1967.
3 Davies, A. L. and Rogers, K. G., *Modal Choice and the Value of Time*, LGORU Rpt. C143, 1973.
4 Thomas, T. C., and Thompson, G. I., *The Value of Time for Commuting Motorists as a Function of their Income Level and Amount of Time Saved*, Highway Res. Record, no. 314, Highway Res. Board, Washington, D.C., 1970.
5 Lee, N. and Dalvi, M. Q., Variations in the Value of Travel Time, *Manchester School*, 37, No. 3, pp. 213–36, 1969.
6 Lee, N. and Dalvi, M. Q., Variations in the Value of Travel Time: Further Analysis, *Manchester School*, 39, no. 3, pp. 187–204, 1971.
7 Heggie, I. G., *Transport Engineering Economics*, McGraw-Hill, 1972.
8 Patterson, J. M., *Factorial urban ecology of Greater Vancouver: Investigations into characteristics of the data base*, M.A. thesis, U.B.C., unpublished, 1973.
9 Rogers, K. G., Townsend, G. M. and Metcalf, A. E., *Planning for the Work Journey*, LGORU, Rpt. C67, 1970.

Many people contributed to this project. Mrs. Ali Metwally acted as research assistant and did most of the data processing; Jim Forbes advised on the layout of the questionnaires; Ken Denike went to a great deal of trouble to prepare indices of household and other socio-economic characteristics for my sample areas; while Lorne Doll, W. G. Waters II, and Peter Jones gave generous criticism and advice. Only I am to blame for the errors that remain.

2

Modal Choice Behaviour and the Value of Travel Time: Recent Empirical Evidence*

by J. H. Earp, R. D. Hall, and M. McDonald

2.1. INTRODUCTION

The choice between competing transport modes, usually known as modal choice or modal split, involves a complex decision related not only to the characteristics of the competing modes, but also to those of the traveller and the trip. Normally a trip is considered in terms of purpose, which may in turn be related to length measured either as time, distance, or generalized cost. The characteristics of the traveller may relate to physical aspects such as age or to social and economic factors such as socio-economic status and income. The modal characteristics include journey time, cost, comfort, convenience, reliability, safety, route choice, scheduling, and interchange facilities.

In addition, an individual may decide not to travel if the journey is unattractive and/or perhaps not essential. This restraint may be introduced artificially (e.g. as part of a policy of restraint) into a transport system and may be considered as an extension of the basic mode choice decision.

The ability to accurately estimate modal split, i.e. the usage of competing modes, is a key ingredient in any transport planning process. It is often done by means of 'diversion curves' which take account of time differences, cost differences, or some generalized time—cost relationship. The model is then calibrated empirically using those aspects of time and cost which are assumed to play a major role in mode choice decisions. As the relative importance of the different factors is likely to change through time, it is essential that correct and accurate relationships be developed. Factors which may appear relatively unimportant under

* this chapter has been published with the permission of the Transport and Road Research Laboratory, Department of the Environment, and the Department of Industry. Any views expressed are not necessarily those of the above organizations.

present circumstances may be crucial to the accurate assessment of future modal choice decisions.

The value of time is generally determined by considering competing modes with alternative journey times and costs. Implied values of comfort, convenience, safety, etc., which also constitute dimensions of 'attractiveness' are thus also included in this value of time. In order to determine the impact of a new or improved mode of transport it is essential that the value of time be related to the new 'attractiveness' factors associated with that mode. In this chapter, the points raised above will be discussed in relation to a series of studies carried out by the Transportation Research Group, Department of Civil Engineering, University of Southampton. The studies reported here include:
(a) Inter-City Studies (1,2) (b) Solent Travel Study (3) and
(c) Bitterne Bus Priority Scheme (4)

Although a major part of this work has been related to the determination of values of travel time and the calculation of an appropriate modal split, all but one of these studies were concerned primarily with more general transport issues. A brief description of each study is given below, together with a summary of those results which have some bearing on our understanding of the modal split process and the value of travel time. The conclusions in section 2.5. relate to all the studies.

The results have cast doubt upon some of the basic concepts and assumptions which have been used in the theoretical framework for determining modal split and the value of time and have thus questioned some of the fundamentals of the way in which economic priorities are assessed.

2.2. INTER-CITY STUDIES

2.2.1. INTRODUCTION

In 1967 the Ministry of Technology, as it was then, commissioned a two-year study to investigate the feasibility of introducing a new form of aircraft on inter-city services within the U.K. The object of this proposed new service was to achieve a reduction in inter-city journey time, albeit at increased fare levels over existing modes. It was postulated that travellers making use of such a service would be undertaking business journeys, and so it was decided to concentrate the studies of travel demand on that element of the market.

In 1969 a further study was undertaken to extend the investigation to cover business travel in Western Europe. In this summary of the work, results from the first stage are referred to as U.K. Results, and those from the second stage as European Results.

In view of the difficulties involved in assembling screen line data based on 'in-transit' inter-city business trips in an unbiased way for all modes, it was decided to conduct the surveys on a 'home interview' type of basis. A number of companies were contacted and given two types of questionnaire, a company questionnaire which related to company policy, the company structure, and over-all trip-making rates as recorded by the company, and an individual questionnaire which related to details of trips made by an employee of the company over a one-month period. This individual questionnaire was completed by a sample of persons within the company, the size of the sample being related to the size of company.

2.2.2. SUMMARY OF RESULTS

Companies clearly placed a value on minimizing journey time for their employees, although the actual value varied with the status of the employee (and hence his level of remuneration) within the company. 'Minimum journey time' was given by the travellers as the most import-ant reason for choosing a mode, although there were indications that they assigned different values to their time during working, leisure, and sleeping periods.

Most employees in the sample had a car available for business journeys, although the use of a car was not necessarily related to the provision of a car by the company.

The total number of days spend each month by individuals under-taking business journeys did not vary significantly between the study areas, although there was a considerable variation in the actual number of trips. This indicated that there was some notional time budget with upper limits and that businessmen appear to allocate a fixed proportion of their time away from the office.

Modal split is related to trip length and, in general, there seem to be distinct distances over which road, rail, and air modes predominate. In conditions of equal time for the over-all trip, travellers prefer car to rail and rail to air. This seems to indicate that despite company policy and individual preference for shorter journey times there are residual factors influencing choice in terms of cost and convenience. The local transport networks and the location of terminals also play an important part in mode choice.

Business travellers seem to have thresholds with respect to both the over-all journey time and the time difference between modes.

The implied values of travel time saving show that there is a considerable difference in the values that individuals place on travel time savings accord-ing to route, mode of travel, and income levels, and that possibly the values are lower for longer inter-city journeys.

2.2.3. BACKGROUND TO RESULTS

Trip motivation for business purposes is extremely complex. The objectives of a business trip can sometimes be achieved by the use of telecommunications, or by the delegation of the task to another individual. Trips can be postponed until either the necessity is more pressing or there are other reasons for travelling and a single journey can accomplish multiple objectives.

The manner of execution of the trip likewise offers a wide range of possibilities, including mode of travel, timing, and itinerary. Each of these factors may be subject to constraints e.g. the mode of travel to and from the terminal, baggage, timing of arrival at destination, frequency or reliability of service, safety, and security. In addition, the range of mode choice for business travellers can be limited by a number of other factors which include the policy of the company, the status of the individual, mode availability, and personal preferences of comfort or service. The costs of business trips are borne by the company, and its travel policy tends to limit the traveller's choice of mode. For example, a businessman who is permitted for a particular journey to travel by first-class rail, though not by air, would not normally value his time or convenience highly enough to be willing to pay the excess air fare himself, although he might choose to travel by second-class rail in which case he would be valuing his comfort or convenience at a lower rate than his company. Company travel policy and individual choice criteria are therefore related but not necessarily identical.

Company Travel Policy

A large number of the firms surveyed did not have a definite travel policy (Table 2.1), although the practice in this respect did seem to be

TABLE 2.1 *Relationship between firm size and occurrence of travel policy*

Firm size Europe UK		Less than 100	100– 500	Greater than 500
Companies having a defined travel policy	(%)	25 66	38 79	55 90
Companies *not* having a defined travel policy	(%)	75 34	62 21	45 10
Number of Companies		138 235	167 255	129 193

related to company size. An analysis of the policies (Table 2.2) clearly showed that minimum over-all journey time was the most important factor affecting mode choice, although its importance varied with category of employee. The table is consistent with the view that the

TABLE 2.2 Company travel policy

Policy (Europe/domestic trips) Europe/International trips United Kingdom trips	Managers	Professional Workers	Intermediate Non-Manual Workers	Junior Non-Manual Foremen and Supervisors	Manual Workers	Total
Permitted modes of travel						
A + RI + RII (%)	(84·5) 92·3 *81·1*	(80·3) 86·2 *54·8*	(46·0) 52·8 *14·2*	(18·2) 27·1 *5·3*	(9·0) 14·5 *2·3*	(44·8) 52·7 *28·4*
RI + RII (%)	(12·4) 6·4 *14·2*	(17·6) 11·5 *15·9*	(12·0) 9·6 *6·7*	(10·1) 9·3 *1·6*	(4·2) 10·3 *1·8*	(11·0) 9·7 *7·2*
A + RII (%)	(4·0) 0·0 *1·9*	(0·0) 0·0 *16·2*	(16·7) 19·2 *27·1*	(24·3) 28·0 *22·8*	(20·1) 21·4 *20·1*	(13·4) 14·4 *17·7*
RII (%)	(2·1) 1·3 *2·8*	(2·1) 1·3 *13·1*	(25·3) 18·4 *52·0*	(47·4) 35·6 *69·2*	(66·7) 53·8 *75·8*	(30·8) 23·2 *46·7*
Factors affecting mode choice						
Min. over-all journey time (%)	(88·4) 91·7 *87·7*	(79·6) 84·9 *80·4*	(57·1) 65·3 *49·0*	(41·7) 54·1 *35·0*	(44·8) 52·2 *26·5*	(64·0) 71·2 *52·9*
Min. cost of journey (%)	(3·6) 1·5 *5·0*	(7·8) 5·2 *13·6*	(26·7) 20·1 *43·9*	(40·2) 31·5 *59·2*	(39·5) 32·5 *66·4*	(22·1) 16·8 *40·6*
Time and cost (%)	(4·0) 3·4 *4·6*	(6·7) 6·9 *3·2*	(10·1) 11·1 *4·8*	(9·6) 8·3 *4·0*	(9·0) 9·2 *5·3*	(7·8) 7·7 *4·4*
Others (%)	(3·1) 3·4 *2·7*	(5·9) 3·0 *2·8*	(6·1) 3·5 *2·3*	(8·5) 6·1 *1·8*	(6·7) 6·1 *1·8*	(6·1) 4·3 *2·1*

A = Air; RI = First Class rail; RII = Second Class rail

value that companies place on travel time savings varies with employee status (level of remuneration). The time of managers is valued highly (90 per cent of the companies gave 'minimum over-all journey time' as the most important factor affecting mode choice), and they are allowed a free choice between available modes. The time of manual workers was considered rather less important, but was still the most important factor in over one third of the companies, although restrictions placed on permitted modes of transport suggest that the actual value placed on their time is relatively small.

Mode Availability

The ownership or more correctly the availability, of a car can affect mode choice in two ways, namely:
(a) as providing an option for use as the main mode,
(b) as a means of access to the main mode terminal
Although the proportion of people with a company car varied significantly between the study areas (Table 2.3), the provision of a company

TABLE 2.3   *Car ownership*

Study area	Company car supplied %	Own car available %	No car available %
Cologne	17·7	74·9	7·4
Copenhagen	25·2	60·9	13·9
Frankfurt	23·6	66·7	9·7
Milan	14·4	80·0	5·6
Zurich	11·0	81·0	8·0
All United Kingdom cities	25·9	65·4	8·7

car does not necessarily imply that it will be used for inter-city business trips, since the marginal cost of running a car (fuel, servicing, wear) is similar to the cost of rail travel (excluding any values attributed to time, comfort, convenience, etc.). An increase in occupancy rates for cars, however, tends to make road travel much cheaper than rail on a *per capita* basis.

Mode Choice Criteria

Within the constraints imposed by company policy and mode availability a traveller will exercise his own judgement in his selection of an appropriate travel mode. The analysis (Table 2.4) of the businessman's reasons for choice of mode showed that 'minimum journey time' was considered to be most important. As would be expected for business trips, the travellers also considered 'fare economy' to be relatively

TABLE 2.4 *Reasons for choice of mode*

Reason for choice	Main mode (%) in European cities *Main mode (%) in U.K. cities*			
	Air	Rail	Road	Ship
Fare economy	3·0	14·3	3·6	4·1
	1·1	*5·7*	*2·8*	
Minimum journey time	82·9	19·6	39·2	22·8
	79·8	*45·7*	*37·3*	
Desire for comfort	2·0	11·7	3·6	2·0
	2·5	*10·3*	*1·1*	
Reliability of service	1·0	10·6	2·7	3·1
	1·4	*9·7*	*2·8*	
Convenience of termini	3·3	7·0	7·3	6·7
	8·9	*14·2*	*3·6*	
Baggage considerations	0·3	1·0	6·2	3·3
	0·1	*0·4*	*3·4*	
Door-to-door convenience	1·2	0·5	12·1	13·7
	3·3	*2·1*	*36·0*	
Require car at destination	1·3	1·4	12·9	22·1
	0·4	*0·6*	*10·9*	
Frequency of service	2·3	5·9	1·2	7·9
	1·1	*3·9*	*0·4*	
No usable connections	1·5	10·7	6·3	0·4
	_*	_*	_*	
Safety	0·3	9·0	0·2	1·0
	_*	_*	_*	
Others	0·9	8·2	4·6	13·0
	1·4	*7·4*	*1·7*	
Total	100·0	100·0	100·0	100·0
	100·0	*100·0*	*100·0*	

* Not considered separately in U.K. questionnaire

unimportant. In the case of 'sleeper' trips in the U.K. the main consideration (23 per cent) was 'reliability of service', whilst 'no loss of working hours' accounted for 8 per cent of these trips. This suggests that the individual traveller assigns a different value to his time during working, leisure, and sleeping periods.

In Europe, the fact that 9 per cent of the travellers cited safety as a reason for choosing rail was probably due to the incidence of aircraft hi-jacking in the months immediately preceding the survey. This issue highlights one of the central and most difficult problems associated with travel choice behaviour. Businessmen often choose a particular mode by rejecting the alternatives. A seemingly 'positive' decision may thus be based upon 'negative' exclusions.

Generation

Businessmen have a certain amount of time available for work each month. Some business trips are considered as essential; others are not and may be replaced by a letter or telephone call. The degree to which less essential trips are made seems to depend upon whether the journey can be completed in a single day. This is confirmed in both the U.K. and European studies, which show that the average businessman spends roughly the same number of days per month on business journeys in each study area, in spite of quite wide variations in the number of trips made. In the U.K. study, for areas far from London the proportion of journeys including an overnight stay was found to be substantially greater than the average for all the study areas as a whole.

Modal Split

The relationship between modal split and trip length is shown in Figure 2.1. It seems to be related to mode availability, over-all trip time, and cost. Three distinct divisions can be discerned: up to 200 km road travel predominates, between 200 km and 400 km rail travel, and beyond 400 km air travel. Although a fundamentally similar pattern to that found in the United Kingdom can be seen in all the countries (Figure 2.1), there are significant differences for each country, which seem to arise from the characteristics and development of the respective transport systems. Figure 2.2 shows the general trend to choose the higher cost rail option as the length of journey increases, implying a willingness to pay for increased comfort under certain conditions. For a selection of major inter-city routes, the trip times for each mode were computed with respect to scheduled service times, making allowance for Trans-European Express Services and first-class rail differences, check-in time and also link trip times to and from the city centres which were taken as representing the centres of demand. The relationship between modal split and time saving for these routes was then plotted (Figure 2.3) and clearly shows the sensitivity of the market to changes in time savings below 3 to 4 hours. Greater time savings seem to have little effect. For this type of trip, the 3 to 4 hours' time saving may be regarded as the threshold above which businessmen usually travel by air. This threshold level may also be related to the traveller's ability to make a complete return journey in a single day, since trip rates seem to increase when journey times are reduced (Figure 2.4). This increase is substantial for time savings under about 5 hours, but if the journey

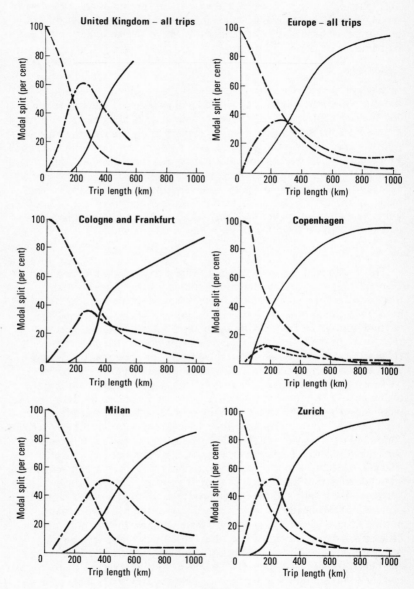

Fig. 2.1

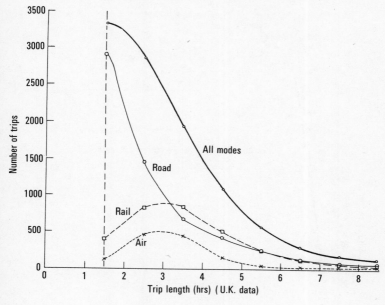

Fig. 2.2

length is greater, i.e. more than can comfortably be managed in a day, the difference disappears.

Value of Time

Implied values of travel time savings were calculated from a knowledge of the time and cost of the journey by alternative modes. For most inter-city trips, the businessman is faced with a number of possible choices. The major dimensions of choice are usually assumed to be relative time and cost, and it is clear from both company travel policies and the individual reasons given for choice of mode that the businessman and his company regard time as the most important single dimension. It is equally clear that, even within the constraints of company travel policy, there are many other factors which affect the decision (including comfort and convenience).

The relationship between mode choice and the value of time depends upon the purpose of the journey, and on the way in which this time can be spent. Even when the traveller is trying to minimize his time, the actual value that he places on it is confounded by the difference between his perceptions and the real time and cost of the alternative modes.

When rail is the fastest mode, assessing the value of time is again

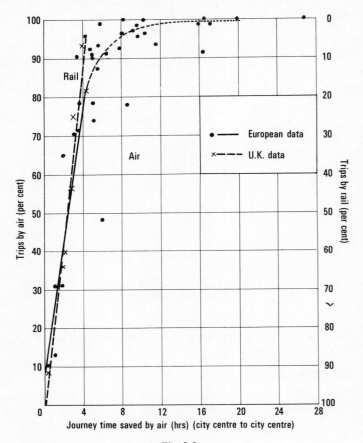

Fig. 2.3

complicated by the choice between first- and second-class tickets, which
does not affect journey time but which does affect the cost. In Europe,
the T.E.E. network introduces an additional choice by offering higher
speeds at a higher price, together with more comfort. In view of the
difficulty of valuing time savings, together with the problems posed by
the type of survey techniques employed, it was not thought appropriate
to ask respondents detailed questions specifically related to the value
of time. An attempt was nevertheless made to develop a single
all-embracing value by considering performance characteristics on forty
routes in Europe.

The differences are identified in Table 2.5 and, in general, reflect

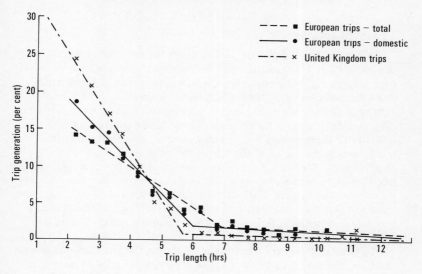

Fig. 2.4

differences in the cost of living and salary scales for senior personnel in the countries involved. A second analysis carried out separately for routes greater or less than 250 km suggests that higher values are associated with shorter trips. It also appears from this distribution that the use of a single 'value of time' does not adequately reflect the wide variation encountered in practice.

TABLE 2.5   *Value of time : by study areas*

City	Value of time saving (£/h)	
	For 50% of businessmen	*For 90% of businessmen*
Cologne	1·50	0·68
Copenhagen	1·15	0·80
Frankfurt	1·60	0·70
Milan	1·00	0·72
Zurich	1·70	0·88
United Kingdom (1968)	0·90	0·62
All Europe	1·20	0·70
European routes <250 km	1·35	0·70
European routes >250 km	1·00	0·76

European results relate to 1970, U.K. results to 1968

2.3 THE SOLENT TRAVEL SURVEY

2.3.1. INTRODUCTION

This study formed part of a series of six studies commissioned by the Ministry of Transport in 1969 to investigate the valuation of travel time savings. Its prime objective was to examine the choice of mode and values of time associated with leisure journeys.

Although there are only a limited number of instances when travellers are (i) making a real choice between one mode and another, and (ii) the majority of them are making leisure journeys, the traffic crossing the Solent to the Isle of Wight during the summer fulfils both these requirements. Because of the 'resort' nature of the Isle of Wight, the majority of travellers crossing the Solent in summer are either holiday-makers or day trippers. They can choose between high-speed, high-cost transport (on the hovercraft and hydrofoil services) and low-speed, low-cost conventional ferries on both the Southampton—Cowes and the Portsmouth—Ryde corridors.

A survey of travellers was conducted during the peak holiday season of 1970. The services and their operating characteristics are shown in Table 2.6, with the routes illustrated in Figure 2.5. Data was collected mainly 'in transit' by direct interview, with additional information

TABLE 2.6 *Characteristics of cross Solent services* (*summer period 1970*)

Route	Service	Travel time	Single fare	Frequency	Comments
Southsea—Ryde	Hovertravel (hovercraft)	7 min	50p	23 crossings/day 09.00—20.00 hrs	40p fare before 10.30 a.m.
Portsmouth—Ryde	Seaspeed (hovercraft)	10 min	39p	18 crossings/day 07.45—19.35 hrs	17 crossings on Sundays
Portsmouth—Ryde	British Rail (ferry)	30 min	29p	21 crossings/day 06.10—23.15 hrs	23 crossings on Saturdays; late service on Fridays
Southampton—Cowes	Seaspeed (hovercraft)	20 min	75p	25 crossings/day 07.30—20.20 hrs	100p day return
Southampton—Cowes	Red Funnel (hydrofoil)	20 min	63p	11 crossings/day 08.00—20.30 hrs	
Southampton—Cowes	Red Funnel (ferry)	55 min	50p	11 crossings/day 06.55—21.55 hrs	60p day return

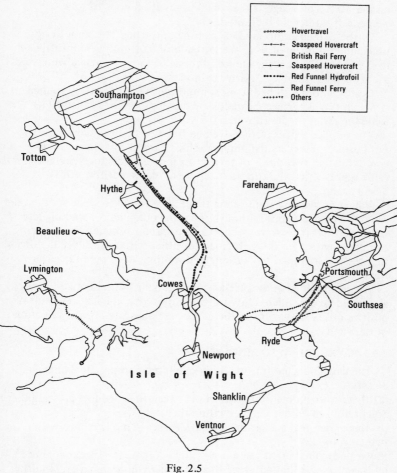

Fig. 2.5

coming from reply-paid questionnaires. The total sample size was over 4900.

2.3.2. RESULTS

Although many travellers were aware of the existence of alternative services, they had in general too little detailed knowledge to make a rational choice based on relative time and cost differences.

A large number of travellers stated that their trip would be either cancelled or postponed if the chosen mode of travel were not available on that day. This suggests that the journey itself constitutes a large element of 'utility' in their mode choice.

Of the two corridors studied, Southampton and Portsmouth, the time and cost differences in the Portsmouth corridor were insufficient to significantly affect the choice of mode of those travellers who were thought to be making a positive choice.

Cost was found to be the most important factor for ferry passengers, although time savings appeared to be more important for travellers using the faster modes (hovercraft/hydrofoil).

The implied values of time seemed to be related to income levels. For business travellers the implied value of travel time savings was approximately two-thirds of their gross income. This unexpectedly low value suggests that the trip itself represents some positive utility to the businessman.

Values of walking and waiting times were found to be larger than the value of in-vehicle time, although the value obtained for walking and waiting for business travel seems to be disproportionately high when compared with the values obtained for other purposes. It was significant that the lowest values for walking and waiting times were for leisure travel.

A large proportion of travellers were making non-essential journeys and could be classified as 'joy riders'. In this case, the large amount of positive utility derived from the mode precludes any normal valuation of time and cost.

2.3.3. BACKGROUND TO RESULTS

The data collected for this study exhibited two principal characteristics:

(a) there were no real route choices between the corridors crossing the Solent but only mode choices within them;

(b) there were real and significant differences in mode choice within the two corridors.

It was unfortunate that the range of time-cost trade-offs experienced by travellers was limited, since this is usually considered a prerequisite for the successful measurement of the value of time. In addition, the novelty of the hovercraft and hydrofoil services also seemed to significantly affect the choice of mode. It was felt that these trips did not provide a valid basis for determining the value of time for leisure travel since they were far too 'mode orientated'.

Mode Choice

The reasons given for the choice of mode (Table 2.7) suggest that relative times and costs do affect the choice between fast and slow modes, but that other factors are also important in explaining displayed

TABLE 2.7   *Reasons for choice of service*

Reasons Stated		Percentage of passengers giving reason					
		Portsmouth–Ryde			Southampton–Cowes		
		Hovertravel	Seaspeed	Ferry	Seaspeed	Hydrofoil	Ferry
Crossing time	(1)*	27	23	3	30	29	5
	(2)	19	15	3	10	17	4
	(3)	6	5	3	3	2	2
Novelty	(1)	25	38	2	24	21	3
	(2)	3	5	–	6	7	2
	(3)	6	5	1	4	4	1
Shortest total travel time	(1)	20	17	–	23	7	–
	(2)	21	22	–	12	12	–
	(3)	10	12	–	3	2	–
Convenient departure or arrival time	(1)	7	7	18	10	14	12
	(2)	10	13	17	11	10	7
	(3)	6	10	8	2	4	2
Comfort	(1)	8	7	14	2	9	18
	(2)	8	6	8	4	10	9
	(3)	12	12	11	3	5	5
Most convenient terminal	(1)	6	4	16	3	5	8
	(2)	15	12	27	8	8	10
	(3)	14	12	13	5	7	6
Most reliable service	(1)	2	–	9	1	–	1
	(2)	4	1	7	3	–	3
	(3)	8	2	13	2	–	6
Availability of parking space	(1)	1	1	–	1	1	1
	(2)	3	–	–	3	1	1
	(3)	7	1	–	3	1	1
Cheapest crossing	(1)	–	–	24	–	–	31
	(2)	–	–	15	–	–	12
	(3)	–	–	8	–	–	4
Fresh air/ sea trip	(1)	–	–	4	–	–	8
	(2)	–	–	1	–	–	4
	(3)	–	–	1	–	–	2
Availability of refreshments	(1)	–	–	2	–	–	2
	(2)	–	–	6	–	–	9
	(3)	–	–	9	–	–	6
Other reasons	(1)	2	1	5	3	2	8
	(2)	1	3	5	1	2	4
	(3)	4	2	7	2	5	4
No answer	(1)	2	2	3	3	2	4
	(2)	16	23	11	42	34	35
	(3)	27	39	26	73	70	62
Total		100	100	100	100	100	100

* (1) most important reason   (2) second reason   (3) third reason

choices. It was evident that 'novelty' aspects influenced the decision to travel by high-speed modes, and, conversely that people travelling by conventional ferry perceived benefits derived from factors of comfort inherent in that form of transport. The results shown in Table 2.7 clearly demonstrate that it is an over simplification to suppose that mode choice is dictated by considerations of time and cost alone.

Any assessment of modal split characteristics implies that travellers have some knowledge of the alternative modes available. The survey asked passengers to state which services between the mainland and the island they knew about before setting out on their journey. Travellers clearly knew about the service on which they were travelling, and the large majority of passengers also knew of the alternative opportunities within the same corridor. At Portsmouth, however, only about one-third were aware of the services between Southampton and Cowes, while at Southampton only about half the passengers were aware of the services between Portsmouth and Ryde. This is probably attributable to the higher proportion of holidaymakers using this corridor. The users of the Hovertravel service showed the least knowledge of alternatives. After the respondents' stated reason for choosing one particular service had been established and their knowledge of alternative ways of crossing the Solent examined, they were asked what they would have done had they known before leaving home that their chosen service was unavailable; e.g. would they have travelled by another service or would they have cancelled or postponed the journey (Table 2.8)? Forty per cent

TABLE 2.8   *Action if service unavailable*

	Percentage of passengers					
	Portsmouth–Ryde			*Southampton–Cowes*		
Action	*Hovertravel*	*Seaspeed*	*Ferry*	*Seaspeed*	*Hydrofoil*	*Ferry*
Cancelled/postponed	46	22	40	20	12	40
Used alternative mode	54	78	60	80	88	60
Total	100	100	100	100	100	100

of the people using conventional ferry services would have cancelled or postponed their journey. Users of the fast modes, however, were more concerned to make their journey at the time planned, and over three-quarters of them (with the exception of Hovertravel passengers) would have used an alternative mode. An assessment of the alternative modes clearly suggests that the great majority of passengers are, for any given journey, orientated to one particular corridor. The small amount of possible route changing thus precluded any further study of time values by examining choices between corridors.

Respondents who said that they would change modes were then

questioned about their knowledge of the fares and frequency of alternative modes (Table 2.9). The accuracy of knowledge of alternative fare levels or frequency was not high among users of any service. Knowledge of the service frequency was generally less accurate than knowledge of the alternative fare. Less than one-third of the sample knew the frequency of the alternative they specified, although passengers using the Portsmouth corridor were slightly more accurate and aware of alternative services than those in the Southampton Corridor.

TABLE 2.9 *Travellers' knowledge of alternative modes*

| | Percentage of travellers knowing fare level and frequency of alternative mode | | | | | |
| | Portsmouth–Ryde | | | Southampton–Cowes | | |
	Hovertravel	Seaspeed	Ferry	Seaspeed	Hydrofoil	Ferry
Fare level						
Correct	20	43	55	45	48	31
Within ± 5p	12	13	8	6	7	4
Within ± 10p	10	7	8	5	7	8
Less accurate/ don't know	58	37	29	44	38	57
Frequency						
Correct	30	45	20	19	25	27
Within ± 10 min	9	6	13	2	4	3
Within ± 20 min	8	5	9	3	6	3
Less accurate/ don't know	53	44	58	76	65	67

Value of Time

In Portsmouth the potential cost and time savings are approximately half those available in the Southampton corridor. For both corridors the terminals at each end are in close proximity, so that the costs of access and egress for both the chosen and the alternative modes are only marginally different. The principal cost and time differences arise in the crossing of the Solent. In the Portsmouth corridor the small cost difference did not seem to be sufficient to deter any particular group of travellers classified either in terms of income or social class. The distribution between modes was not significantly different and clearly indicated that users did not discriminate between the services in terms of either cost or time. Other factors, e.g. novelty, seemed to play an important part in the choice of mode in this corridor, so that any analysis of mode choice seemed unlikely to lead to any improvement in mode choice models or to render significant values of travel time.

In the Southampton corridor, differences in cost and time represented a greater proportion of the total journey cost. Total travel times and costs were also greater, in absolute terms, than those experienced in the Portsmouth—Ryde corridor. Fewer people, therefore, used the services purely for pleasure, i.e. 'joy riding', so that the information obtained from the sample of passengers using this corridor led to more significant values of time. A higher proportion of high-income passengers used fast modes, implying that their mode choice was related to the shorter travel time. The fact that multiple choices were possible in both corridors, with two 'fast' modes and one 'slow' mode, made it difficult for the passengers to make rational choices. Incomplete knowledge of the frequency and cost of alternative modes was not surprising in this complicated situation. The actual timetables led to further difficulties. Variations in departure times made it extremely hard to establish accurately how much time would be saved by choosing one mode rather than another. On some occasions the slow mode departed ten minutes earlier than the fast mode, thereby reducing the inherent advantage of the fast mode. Such difficulties, which were partly due to timetabling, were extremely complicated (and may well account for the poor knowledge of frequency), making it unreasonable to assume, particularly for leisure or holiday travel, that users were aware of all the opportunities available to them.

These data limitations made it difficult to develop accurate mode choice models using the techniques of multivariate statistical analysis. The analysis was carried out in four stages, each stage reflecting a change in emphasis. They included a consideration of the scope of data, the corridor of travel, the form of the independent variables, and the type of multivariate statistical function being used. Each of these phases is summarized in Table 2.10.

A summary of time values obtained from each phase of the analysis for each of the major purposes of travel is shown in Table 2.11. It was not possible to obtain any meaningful answers from the data collected in the Portsmouth—Ryde corridor, so that the analysis of this corridor was terminated at the end of phase 2. Some difficulty was experienced in obtaining 'notionally' correct signs on the coefficients related to differences of times and costs, and this is shown by the use of brackets in Table 2.11.

The results from the Southampton—Cowes corridor which had the correct signs on the coefficients (giving more plausible values of time) gave a maximum, for all trip purposes, of £0·81 for business travel in phase 2b. As would be expected, the values obtained for walking and waiting times were significantly higher. Moreover, the values of time which were related to income in phase 4b showed the highest level for business travel and the lowest for holiday travel.

The results from phase 3, which were derived from the PROLO programme (using multiple probit and logit analysis) were not signifi-

TABLE 2.10 *Summary of analysis*

Phase of analysis	Data	Corridor	Model type	Variable form
1 (a)	Basic (unweighted)	Portsmouth Southampton Combined	Discriminant Initial	Time and cost, absolute values and difference
(b)	Basic (weighted)	Portsmouth Southampton Combined	Discriminant Initial	Time and cost, absolute values and difference
(c)	Basic (unweighted)	Portsmouth Southampton Combined	Discriminant	Time and cost, absolute values and difference Income Car mileage 2p and 5p
2 (a)	Revised (all journeys)	Portsmouth Southampton	Discriminant	Time and cost difference
(b)	Revised (≤25 km)	Portsmouth Southampton	Discriminant	Time and cost difference Income Car mileage 5p
3	Revised (≤25 km)	Southampton	Probit Logit	Time and cost difference Income
4 (a)	Revised (all) (≤25 km)	Southampton	Discriminant	Walk/wait time difference Link mode time difference Cost difference
(b)	Revised (≤25 km)	Southampton	Discriminant	Income and time difference Cost difference

Notes: (i) Basic data: trips were assigned in the model as 'fast' if the mode was hovercraft or hydrofoil and as 'slow' if conventional ferry

Revised data: trips using the chosen or alternative mode were assigned in model as 'fast' or 'slow' if the total trip time using that mode was respectively less than or greater than that using the other mode

(ii) ≤25 km refers to the mainland access or egress distance

cantly different from those derived by multiple descriminant analysis in phase 2b. This suggests that the variability of answers is a function of the data (or, more correctly, of the type of mode choice involved on the Solent) and not of the method of analysis.

The use of absolute times and costs as the independent variable in the initial series of models using descriminant analysis did not lead to any statistically significant values of time. Moreover, the meaning of these answers was not clear, since many of the values for the users of the slow mode were higher than the equivalent values derived for the fast mode. Since this method incorporates no measure of time or cost difference, the answers may reflect the rate of spending through time in relation to the service rather than the value of the actual time saved. As a result, absolute forms of time and cost, together with ratios, were omitted from the remaining analysis, and time and cost differences were used as the major determinants of choice.

TABLE 2.11 *Summary of time values (£/hour) (Based on time and cost differences)*

Phase of analysis	Travel purpose					
	Holiday	Business @ 2p /mile	@ 5p /mile	Other	Pleasure with novelty	without novelty
1a	0·44 *(0·03)*	(2.21) *(0·42)*	(1·48) *(0·77)*	0·41 *(1·73)*	0·01 *(0·13)*	(0·14) *(0·21)*
1b	0·63 *(0·02)*	(2·50) *(0·46)*	– –	0·46 *(0·89)*	(0·01) *(0·10)*	– –
2a	0·10 *(0·02)*	(1·91) *(1·09)*	– –	0·04 *(2·96)*	0·41 *(0·02)*	(0·05) *(0·19)*
2b	(0·10) -	0·71 *(1·07)*	0·81 –	0·12 –	(0·93) *(0·01)*	(0·47) –
3 Probit Logit	(0·04) (0·04)	0·72 0·73	– –	0·12 0·12	(0·93) (0·93)	– –
4a Walk/wait	0·58	–	3·10	0·76*	–	0·25
4b % gross income	14	–	67	27*	–	53

Figures in italics refer to Portsmouth corridor, others to Southampton corridor. A value in brackets is formed from time and cost difference coefficients having one or both signs opposite to the expected signs

    * Significant at 95% level

It is clear from the analysis in phase 4a that the values of time derived using walking and waiting times as the independent variables are generally higher than those obtained using time and cost differences for the over-all journey. This follows the general trend observed in other studies which show that time spent walking and waiting is valued more highly than in-vehicle time, although the value obtained for business travel seems to be disproportionately high compared to those obtained for other purposes. It may well be significant that the lowest value of time derived in this way relates to the models of leisure travel (£0·25/h). Using income as an independent variable was always significant when used in the analysis of the Southampton corridor data. The results showed that in travel associated with business the traveller valued his time at approximately two-thirds of his gross income rate. Non-business travel gave lower values of time, the average being approximately one-

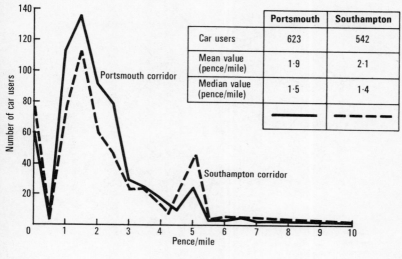

Fig. 2.6

third of gross income rate. The results for business travel are shown with mileage costs for car access at 2p per mile and 5p per mile (Figure 2.6).

The results obtained in phase 2b of the analysis proved to have the highest 't-test' values although these were not statistically significant. Values of time for 'walking and waiting' exceeded £3 per hour.

2.4. BITTERNE BUS PRIORITY SCHEME

2.4.1. INTRODUCTION

In 1969 the Minister of Transport set up a Working Group on Urban Transport Problems which recommended that a series of Bus Demonstration Projects should be introduced. The Bitterne Bus Priority Scheme (Figure 2.7) was one of these projects. The scheme involved the application of a traffic signal control system to the movement of vehicles on to and along 5 km of the main A3024 radial road through Bitterne, a suburb of Southampton, so that vehicular flow on the Main Road remained in a 'free flow' condition at all times during morning peak periods. In order to achieve this vehicles had to be held back on side roads by controlling access to the Main Road during periods of heavy demand. Control was achieved by creating one-way streets, by banning certain peak-hour turning movements, and by using no-entry signs so that all vehicular traffic was forced to gain access through junctions controlled by traffic signals. The signals were linked together to optimize flow along the Main Road whilst at the same time varying the degree of control

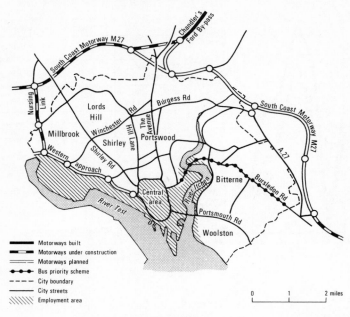

Fig. 2.7

or restraint on side roads in response to conditions prevailing at key locations on the Main Road. Other vehicular traffic crossing the Main Road was directed along specific routes to avoid queues in side roads. Buses were given similar specific routings, using bus-only lanes and bus-only turning movements to facilitate direct access to the Main Road on which they were able to operate in free flow conditions. Delays to buses were thereby minimized by transferring any vehicular traffic delays from the Main Road to side roads.

The survey work, which was in four stages, was principally aimed at measuring the change in total vehicular and person movements attributable to the implementation of the scheme and included:
(a) the monitoring of the change in bus journey times and in the reliability of service measured by any deviations from the regular schedule, bus occupancies, and the waiting time of passengers at stops. The monitoring of buses and passengers was undertaken by two series of surveys--on-bus monitoring and bus-stop monitoring;
(b) screen line flows over the River Itchen bridges;
(c) vehicle speeds on the main roads and side roads;
(d) turning movements and vehicular delays at junctions;
(e) a home interview of 10 per cent of households in the area, together

with a roadside interview of a proportion of vehicles (other than buses) entering the area to establish an origin and destination pattern for the area;

(f) other information, including violation of road signs and control mechanisms, accident statistics, unemployment statistics, and trend statistics of bus passengers.

2.4.2. SUMMARY OF RESULTS

The traffic control system created improved conditions of vehicular flow on the Main Road.

Together with the improvement on the Main Road, the measures giving priority of access to buses brought about a real and significant reduction in the journey time of buses and enabled them to adhere more closely to their published schedules.

There was no significant change in the total journey times of other vehicles, although delays to these vehicles were redistributed from the Main Road on to the access roads.

There was no significant change in the distribution of vehicle movements across the River Itchen screen line, although vehicles were compulsorily diverted within the area of the scheme.

The modal split was not significantly altered.

The change in delay patterns to cars meant that car drivers thought that their journey times had increased.

2.4.3. BACKGROUND TO RESULTS

The Bitterne Scheme attempted to assess the feasibility of a Bus Priority Measure in conditions where main road bus lanes could not be introduced. Although it was hoped that an increase in the 'attractiveness' of bus journeys would alter the modal split, it was intended to evaluate the over-all effectiveness of the scheme by measuring changes in person movements. The scheme was intended to manage the traffic (movement of people) rather than to restrain it.

Traffic flow

The most significant result which emerged from monitoring the movement of buses was the decrease in the variation from the published schedule at Northam (from an average of 3·7 minutes late to 0·3 minutes late) together with a reduction in its standard deviation (4·7 minutes to 3·3 minutes). This represents an over-all improvement of 3·5 — 4·0 minutes per passenger and since there was no significant change in either the variation from schedule at approach road bus stops, or in bus waiting times at Main Road bus stops, the improvement is directly attributable to two factors:

(i) an increase in bus speeds on the Main Road due to improved

flow conditions, and
(ii) improved side road access to the Main Road.
Bus speeds along the principal section of the Main Road increased from
16·4 km/hr to 21·7 km/hr, while gains at entry points varied according
to the nature of the entry control mechanism and the length of queue
experienced before the new scheme was introduced. On the Main Road,
the average (non-bus) vehicle speed increased by 25 per cent from
23·7 km/hr to 29·5 km/hr, with a corresponding increase in the side
road delay from 1·9 minutes to 2·8 minutes. The increased efficiency
of the traffic flows can be clearly seen in Figure 2.8. In general, there
were increased delays at the designated entry points and decreases in
delays on cross routes. The River Itchen screen line surveys showed
that there were no major changes in corridor movements within the
area, nor any significant changes in car occupancy rates (Table 2.12).

TABLE 2.12   *Car occupancy rates at Northam Bridge (persons/car)*

Inbound (07·00−09·30)			Outbound (16·00−18·30)	
Before	After (1st study)	After (2nd study)	Before	After
1·34	1·32	1·30	1·49	1·45

The minor changes in the patterns of vehicular flow over Northam
Bridge, and of those entering or crossing the Main Road, clearly result
from the compulsory control measures.

Modal Split

The modal split remained largely unchanged (Table 2.13), and the
noticeable increase in bus patronage during the first few months after
the introduction of the scheme was probably due to added publicity.
Spare capacity was always available even during this period of increased
patronage.

The scheme was such that priority could only be given to buses in
the morning peak. The advantages were therefore only operative for half
the daily commuting journey, there being no significant difference in
journey times during the evening peak. In addition the home interview
survey showed that the average journey time by bus was 31 minutes
while the average journey time by car was 23 minutes. Since these figures
relate to the 'after' survey it is clear that the car still provided the
shorter journey time.

Perception of Time

The home interviews included a section on the perceived change in
bus journey times. These are summarized in Table 2.14. For passengers
travelling along the Main Road across the River Itchen the average

TABLE 2.13   *Change in mode: 'Before and After' Studies*

Mode used after experiment	Mode used before experiment %							
	Car driver	Car passenger	Motor cycle	Bus	Walk	Others	Total	Sample size
Home interview								
Car driver	33·1	0·3	0·2	0·4	0·1	0·2	34·5	780
Car passenger	0·8	9·0	0·1	1·0	0·1	0·0	11·0	249
Motor cycle	0·2	0·0	3·6	0·3	0·0	0·0	4·1	93
Bus	0·6	0·2	0·3	25·1	0·2	0·1	26·4	598
Walk	0·1	0·0	0·0	0·3	19·0	0·1	19·5	442
Others	0·2	0·0	0·0	0·1	0·0	4·0	4·4	101
All modes	35·0	9·9	4·2	27·2	19·4	4·6	100·0	2263
Roadside interview								
Car driver and passenger	86·1	10·4	0·9	1·2	0·1	1·3	100·0	2049

TABLE 2.14   *Perceived change in journey time : 'Before and After' Studies*

	Sample size	Perceived change in time					Average time change**
		Less	Same	Greater	No answer	Total	
Bus passengers' journey time (home interview)							
All passengers	563	31·1	47·8	9·9	5·2	100·0	+ 2·87
Passengers crossing Northam Bridge	349	51·6	36·7	6·3	5·4	100·0	+ 4·87
Bus passengers' waiting time (home interview)							
All passengers	563	16·5	64·1	13·0	6·4	100·0	5·61†
Car users' journey time (home interview)							
All users	1090	18·4	51·9	24·0	5·7	100·0	−0·76
Bus experience*	112	27·6	46·3	25·0	0·1	100·0	—
Car users journey time (roadside interview)							
All users	6949	18·6	51·0	27·6	2·8	100·0	−1·29
Bus experience*	802	21·3	52·7	25·7	0·3	100·0	—

* Car users who had experienced the bus service between 'Before and After' Studies
** Answer in minutes (gain + loss —)
† Average perceived waiting time

perceived reduction in bus journey times was found to be 4·9 minutes which compares with a measured reduction of 3·5 minutes.

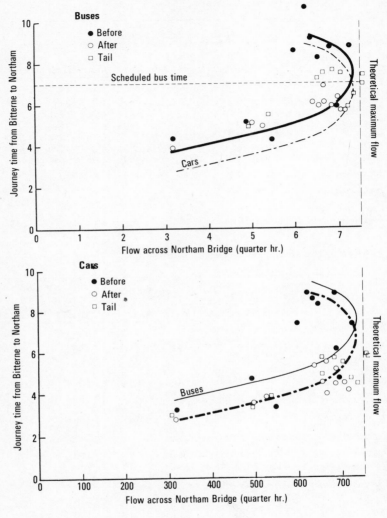

Fig. 2.8

Car travellers perceived an increase in journey time of approximately one minute although the measured times showed that they actually enjoyed a slight reduction in journey time. This perceived increase in journey time is probably accounted for by the difference in delay patterns, i.e. car travellers prefer to move slowly on a congested road rather than quickly on a non-congested road with a stationary delay at the entry point.

2.5. VALUATION OF TIME SAVINGS

As a direct result of inconclusive theoretical work (5) much of the effort to establish acceptable values for time savings has concentrated on empirical travel time studies. Although a wide variety of techniques have been used in an attempt to establish time values (report by Barber and Searle (6) in their review of recent U.K. work) the majority of studies have either used a limiting time value technique, similar to that reported in section 2.4. (Inter-City Studies) or have used discriminant or logit analysis like that adopted in section 2.3. (Solent Travel Study).

Both approaches depend on the assumption that mode choice is based on a trade-off between the relative time and cost of alternative modes and that this trade-off implies a relationship between time and cost reflecting the passengers' perceived 'value of time'. This assumption is based on the fact that time and cost tend to be common factors, if not always the dominant factors, in all mode choice decisions. The use of a time-cost relationship, which implies a value of time, nevertheless ignores the important fact that these two variables may not completely encompass the principal criteria of choice. One of the most striking features of the studies carried out by the Transportation Research Group is the degree to which external factors other than time and cost reflect modal choice decisions. The attractiveness of the mode, e.g. comfort, convenience, and safety, is generally overlooked when formulating values of time or when calculating the modal split. The time—cost approach is therefore too simplistic, particularly when conditions are changing through time. The Solent Travel Study also suggests that many travellers who are normally assumed to be making a choice based on relative times and costs, do not in fact perceive—or even wish to avail themselves of—an alternative mode and would be content to postpone or cancel their trip if the chosen mode were not available. In many seemingly straightforward choice situations the assumption that a real choice exists, even if it is only walking, and that the traveller recognizes all alternatives and bases his choice on a comparison of their relative times and costs, is not valid. In reality, this assumption obscures the fact that the implicit values of time derived from examining mode choice cannot necessarily be applied to the population of travellers as a whole.

A further factor which emerged from the above studies is the wide variation between the real and perceived cost and time elements associated with a journey. It is clear that the assessments of time and cost are influenced by the conditions under which mode choices are made and by differences in the scheduling of alternatives (and are also possibly related to the itinerary of the travellers).

The results suggest that values of time obtained by existing statistical methods must be treated with caution. Three particular aspects of modal choice models which use the concept of generalized time or cost require more detailed consideration. These are:

(a) trip motivation
(b) mode attractiveness
(c) perception of time and cost.

Each aspect is examined in detail in the following sections and there is some evidence from the above studies, together with indications from other research work, to suggest that the existing methods of assessing values of time for use in modal split models could be improved in the following ways:

(a) changing the approach to modal choice to take account of the more fundamental decisions which are made before choosing between available modes for any specific journey;

(b) testing the values obtained from empirical studies in a more rigorous way to ascertain to what extent they are mode specific and therefore cannot be used in situations dissimilar to those prevailing during the experiments;

(c) understanding more about the traveller's perception of time and cost, particularly with respect to the usefulness (or wastefulness) of time spent on the mode or in gaining access or egress, and in the weighting/value of the time saved in relation to the over-all journey or to the time of day, week, or year, in which the saving is obtained.

2.5.1. TRIP MOTIVATION

One of the most significant facts which emerged from the Solent Travel Study was that up to 40 per cent of travellers would cancel or postpone their trip if the chosen mode were not available. In the study of inter-city business trips it was likewise discovered that the total number of days spent on trips away from the office each month was limited, implying an apparent time budget. In both cases the evidence suggested a clear relationship between the characteristics of the journey and the motivation behind it.

Empirical studies of mode choice tend to ignore these motivational aspects. They are simply subsumed within the journey purpose and are assumed constant for each purpose sub-group. For certain journey purposes, e.g. commuting, the motivation may be similar; for others it is likely to vary considerably. Each traveller has a definite amount of time and money available to satisfy his needs. In some cases these needs will have spatial dimensions and will, therefore, involve travel. Motives will therefore, be related to the total time and money resources available to the traveller (i.e. to his time and cost budgets and constraints) and, in turn, will vary depending upon alternative ways in which these resources might be used. Since it is clear that journeys can always be postponed or cancelled (or even substituted for each other) the process of arriving at a journey decision implies two stages of choice. This is shown in Figure 2.9. The two stages or levels of decisions comprising this 'model' are:

(a) *'Strategic Level'* In this the traveller assesses the need to make a journey, i.e. he becomes motivated and uses simple constructs of journey time and cost to evaluate his ability to satisfy this need—with any journey appearing as a derived demand—within the broad constraints of available time and cost. He will then choose either to make the trip (with or without adjusting his over-all budget) or not to make it.

(b) *'Tactical Level'* Having decided on the above strategic basis to make a journey (possibly using average values of time) he then examines the detailed mode and route choices available. During this process he may again refer to his time and cost constraints, and may even reconsider the basic journey need, while examining the potential trade-off between modes in terms of an implied marginal value of time.

Empirical studies of modal split are usually only undertaken at the tactical level. It is tacitly assumed that all the travellers being observed, or those of them who could have used an alternative route, are positively involved in some kind of choice procedure. This assumption ignores the possibility that the sample:

(a) might include travellers who have decided at a strategic level to restrict their choice to a single mode because of time and cost constraints related to other commitments outside the range of the present decision (they might similarly base it on the tactical decision that only one mode is able to satisfy certain overriding constraints of time and/or cost);

(b) might include those people who have made, or had imposed on them, a fundamental decision to restrict the possibility of choice by assigning their disposable time and income to other activities. For example, a businessman may choose air travel not out of preference but because it is the only mode that meets his over-all time budget. A housewife may likewise choose not to travel because she cannot afford to do so within her cost budget; or a commuter may choose to live next to a public transport corridor (and decide not to own a car), thereby restricting his choice in advance of any journey decision. There is some evidence confirming this in the Bitterne Study and also in the work reported by Wabe (7).

These decisions are clearly not dealt with by the current procedure for valuing travel time or for predicting modal split. The travel decision— or its motive—will generally be affected by the individual's knowledge

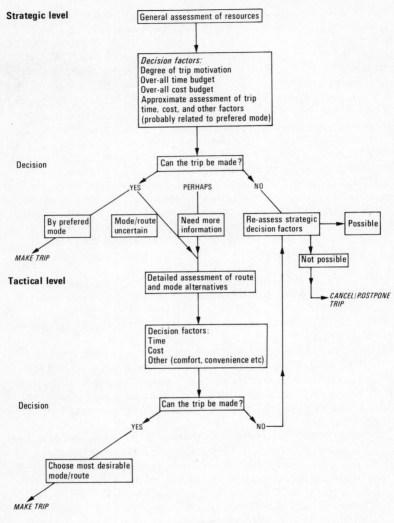

Fig. 2.9

and perception of the degree of the transferability of the journey (i.e. to another person), the opportunities for postponing or cancelling, and the interdependence between himself and other travellers. His over-all pattern of trip making, the frequency of similar trips, and his experience of the physical characteristics of the travel mode will also have some impact on this journey decision. Applying the above criteria to three basic trip purposes—commuting, business travel, and leisure travel—yields totally different profiles of the constraints imposed on mode choice.

A Commuter pays his fare from a personal (possibly household) budget and must generally travel during peak periods. The distance travelled might be restricted, but the frequency of the journey usually means that he has a detailed knowledge of the reliability of service and of most alternative means of travel. The need to make overnight stays or to involve other people in his decisions in low or non-existent. The journey is usually inflexible, although this may change with the introduction of flexible working hours. Car availability tends to be linked to his income and to the joint arrangements for use made with other members of the family.

A Business Traveller will have his fare paid by the company—they may even provide him with a car—but will be restricted by company travel policy. His time constraint is broadly related to the working day or week. Journey length varies considerably, as does frequency, usually in relation to company status. His knowledge of different services will vary according to his general travel pattern, although his assessment of alternative costs and times will probably be reasonably accurate. The trips may well involve overnight stays. Possibly the most important feature of business travel is the degree of flexibility available for postponing, cancelling, substituting (by telecommunication), transferring, or linking the trip (or objective) to another journey or colleague.

A Leisure Traveller (other than for holiday) meets his travel cost from a personal (or household) budget so that trip making is generally constrained to evenings and weekends. The length of trips will normally be of modest duration with a fairly low frequency. It is possible for an overnight stay to be included, and other people, or members of his family, will usually be involved in the trip. Car availability will tend to be higher than for a commuting trip. Most important is the high degree of flexibility in terms of postponing, cancelling, or substituting the objective of the journey.

It is evident from these summaries that the constraints associated with different trip purposes have a fundamental effect on motivation, and hence on the choice of mode, from the earliest stages of the decision. Only people with a high propensity to travel will proceed to the second or tactical stage of modal choice. It is therefore highly probable that empirical studies render biased values of time.

2.5.2 MODE ATTRACTIVENESS

The normal method of estimating a modal choice model assumes that all travel characteristics can be expressed in terms of generalized time or cost. Estimates of time and cost for each element of the journey by each of the alternative modes are first calculated, using appropriate weighting factors to allow for the difference in real and perceived values. Mode choice is then predicted on the basis of variations in this generalized time or cost. The justification for using a 'value of time' rests heavily on the assumption that the other determinants of travel choice can be ignored, either because they have little or no significance or because they are effectively embodied in the value derived empirically.

It is clear from the studies of inter-city business travel, and from the Solent Travel Study, that many travellers do not feel that their choices are based solely on time and cost, but are often conscious of other factors relating to the nature of the services offered in terms of a positive attribute of the chosen mode or of a negative attribute of the rejected mode. These factors could be regarded as measures of 'mode attractiveness'. In the light of these results, and of the difficulties generally experienced in calibrating predictive models, it is clear that more effort should be devoted to assessing what effect factors other than time and cost have on attitudes towards competing modes and what effect these attitudes have on the values of time derived from empirical studies.

Mode attractiveness might include the passenger's perception of:
degree of comfort (overcrowding, seat availability, etc.);
convenience of access;
safety;
reliability;
usefulness of travelling time (i.e. ability to use it for other purposes such as reading);
service frequency;
traveller's itinerary.

Detailed research would indicate which of these factors (e.g. service frequency) emerge as important determinants of choice with a direct time or cost equivalence and should therefore be included in the implied time values obtained in empirical studies. It is unlikely, however, that mode preference could ever be completely specified in terms of time and cost alone. Factors like perceived safety may override all other considerations and give rise to the seemingly irrational choices observed in most modal split studies. A prime example of 'mode attractiveness' is seen in the behaviour of the traveller who chooses first-class rail travel in order to derive a benefit which is clearly not related to any travel time saving. It is therefore, important to note that any values of time obtained from empirical studies should be recognized as being mode specific and should only be used in circumstances comparable to those pertaining during the original experiment.

2.5.3. PERCEPTION OF TIME AND COST

One of the most difficult problems encountered in valuing travel time savings is the discrepancy between real and perceived values of time and cost. Since modal split forecasts must be based on real times and costs, particularly when new modes are concerned, it is of paramount importance to have a clear understanding of the relationship between the real values of time and cost and those perceived by the traveller. There is also evidence to suggest that time savings are not necessarily perceived in a linear manner, but are subject to thresholds related to the size of the saving and to the occasion on which it occurs.

Costs

The perception of travel costs seems to be related both to the purpose and to the mode of travel. Journeys by car seem to be the hardest to evaluate. The Solent Travel Study suggests that there are two main classes of user: those who price the journey in terms of petrol costs alone (or short-run marginal cost), and those who use a 'fully allocated' cost (generally based on the mileage costs reimbursed by a company). It is clear from the extent to which these results are linked to trip purpose that there are fundamental differences in the way that costs are evaluated and that this in turn is reflected in implied value of time savings. This marked difference in approach makes it difficult to assess accurately the costs associated with access and egress by car to inter-city travel modes. Indeed, they may not even be considered as part of the over-all cost of a trip in which air or rail acts as the main mode. More attention should therefore be paid to the perceived values of 'link' mode costs.

Time

The results obtained from the Solent and Bitterne studies confirm that walking and waiting times are valued more highly than the equivalent 'in-mode' values. This is probably because the 'walking and waiting mode' is less attractive than the others, either because it is less comfortable, or because it is accompanied by anxiety about the reliability of the main mode. The difference may also be due to a change in the relationship between perceived and real time which might in turn be related to attractiveness. Comparative studies of real and perceived time values may therefore determine the 'time equivalence' of some aspect of attractiveness. Since perceived values of time are also included in mode choice decisions it is necessary to monitor variations within each trip. In the Bitterne Study, car travellers thought their journeys were longer after the introduction of the scheme, although they had in fact fallen slightly. This was probably caused by a redistribution of existing delays from the over-all journey to a specific delay at the entrance to the main road, and the reason for the increase in perceived time may have been due to the single block of delay and/or to the anxiety caused because it occurred some distance from their ultimate destination.

Values of time may not, therefore, be a linear function of journey length (8,9). This concept is supported by the Solent and Inter-City studies, which show that the values of time seem to be related to cost differences and cost ratios. The use of curvilinear continuous functions seems quite logical, although the above studies, together with work by Sharlach (10), suggest that the function may not be wholly continuous but contain discrete steps related to discrete time periods, e.g. one day. For business travel a step seems to occur after a journey time of five hours, which is probably related to the ability of the traveller to complete the return journey in one day.

The above thresholds will clearly be affected by the opportunity cost of travel time involved, i.e. to its possible use for other purposes. For example, a business traveller who combines a one-hour rail journey with lunch may not place any value on reduced journey time, and indeed a reduction in journey time may have a positive disutility in that it would no longer leave time for lunch. Travellers using the rail sleeper mode in the inter-city study were found to exhibit a similar effect. Sharlach has likewise expanded the concept of the usefulness of time (its opportunity cost) by suggesting that there might be standard patterns throughout the day.

A further factor, illustrated in the Solent Travel Study data, is that changes in travel time may have no over-all effect on mode choice if they are not also associated with change in scheduling.

2.5.4. CONCLUSIONS

1. Mode choice may be considerably affected by the motivation or purpose of the journey as well as by the perceived efficiency and attractiveness of the alternative modes. Existing procedures for estimating modal choice do not take into account these strategic decisions, related to the over-all time or money constraints imposed on the traveller. This tends to bias the estimates of time values derived from empirical studies.

2. Values of time (or implied values) are usually derived empirically by examining situations in which the traveller is supposed to be trading off time against cost. However, although this trade-off may exist in an abstract sense, the traveller does not always perceive the trade-off (or may not know of or consider any alternatives), and this may lead to spurious results.

3. Values of time derived from modal choice situations are mode-specific and should only be applied within the range of data upon which

they are based. The evidence suggests that time values vary in relation to the length of journey, the size of the time savings, and the time of day. A single aggregated value of time is therefore inappropriate for modal split predictions.

4. Values of time are usually derived from a situation in which time is traded against cost. They should not, therefore, be used to assess the benefit of time savings in circumstances where time and cost are not traded against each other.

5. The apparently irrational choice observed in empirical studies suggest that elements of mode attractiveness (giving a more complete description of the behavioural pattern of the trip maker) are not adequately explained by current models. Values of time derived from a time-cost trade-off may therefore embody some measure of the relative attractiveness of the competing modes.

REFERENCES

1 Transportation Research Group, University of Southampton, *Inter–City VTOL Studies (U.K.), Business Traffic* (Vol. 1), Final Report to the Ministry of Technology, 1969.

2 Transportation Research Group, University of Southampton, *VTOL: A European Study, Business Traffic* (Vol. 3), Final Report to the Department of Trade and Industry, 1971.

3 Transportation Research Group, University of Southampton, *The Solent Travel Study,* Final Report to the Department of the Environment, 1974.

4 Transportation Research Group, University of Southampton, *An Evaluation of the Bitterne Bus Priority Scheme, Southampton,* Final Report to the Transport and Road Research Laboratory, 1974.

5 Mansfield, N. W., *The Analysis of Route-Choice and Modal-Split Situations in Research into the Value of time,* Highways Economics Unit, Department of the Environment, 1969.

6 Barber, J. and Searle, G., *The Value of Travel Time Savings–U.K. Studies, Transportation and Environment : Policies, Plans, Practice,* University of Southampton, 1973.

7. Wabe, J. S., A Study of House Prices as a Means of Establishing the Value of Journey Time, the Rate of Time Preference and the Valuation of some Aspects of Environment in the London Metropolitan Region, *Applied Economics,* December 1971.

8 Gronau, R., *The Value of Time in Passenger Transportation : The Demand for Air Travel,* National Bureau of Economic Research, 1970.

9 Hoinville, G. and Johnson, E., *The Importance and Value Commuters Attach to Time Savings,* Social and Community Planning Research, 1971.

10 Sharlach, H., *Report on Internal German Air Services,* Cologne, 1966.

3

The Skyport Special:
An Experimental
Personalized Bus Service*

by Margaret J. Heraty

3.1. INTRODUCTION

The Skyport Special was an experimental pre-booked bus service run by
London Transport for workers at Heathrow Airport in London. The
planning of the service was based upon an extensive journey-to-work
survey of all airport workers which was then analysed by a generalized
cost model and used to forecast the likely levels of patronage. However,
the patronage achieved was substantially lower than that forecast, and
attitudinal research was carried out to investigate the discrepancy. The
research indicated that the assumptions of the model overlooked a
number of factors which were important in modal choice decisions, at
least in so far as the population affected by this service was concerned.

The model was then rerun with some parameters altered to attempt
to take account of the important behavioural factors isolated in the
attitudinal research. This resulted in a new estimate of patronage which
was remarkably close to the measured usage which actually materialized.

Although the experimental service was not in itself an operational
success, the results of the research indicated where classical modal choice
models might be improved and were therefore of value in giving a lead
to future travel choice work, particularly in relation to new and innovatory public transport services.

* This chapter is based, in part, upon a study carried out by Freeman Fox and
Associates for the Operational Research Department of London Transport. The
author is grateful for the permission given by both organizations to publish this
chapter. She wishes to acknowledge the extensive contributions made by
David Day and Val Green of London Transport and Richard Scurfield of Freeman
Fox and Associates to earlier publications referred to in this chapter (References
1, 2, 3, 4, and 5). However, all responsibility for the personal views expressed
rests with the author.

3.2. PLANNING THE SERVICE

3.2.1. BACKGROUND TO THE EXPERIMENT

In the late 1960s a great deal of interest was shown by London Transport (LT) in the possibility of operating services for regular travellers between fairly compact residential areas and centres of employment. The model for most of these 'personalized' services was the Premium Special project in Peoria, U.S.A., which involved running experimental contractual, home-to-work, express services to the Caterpillar Tractor Co. In this country, similar experiments were mounted in Manchester (the Hale Barns Executive Coach Service) and in Stevenage (the Blue Arrow Service).

However, in London most residential areas have fairly good rail services to the Central Area and it was likely that road journey times, even by express coach, would be unattractive without extensive bus priority measures. Also there is a much wider variety of destinations in Central London than in smaller places, so that it would be extremely difficult to provide a personalized service with adequate coverage of the main employment areas.

For these reasons London Transport concluded that this sort of service would have a much greater chance of success if directed to an employment centre outside Central London. Such a centre was Heathrow Airport. The number of employees was large, over 30 000 being on duty on any one day. A large proportion lived in areas close to Heathrow, and the majority travelled to work by private car. The latter factor was important, in that the main aim was to attract traffic away from private cars. These considerations suggested the possibility of a promising level of demand for a personalized service, particularly from south of the Airport where existing public transport facilities were generally inconvenient and private car usage is high.

3.2.2. INITIAL JOURNEY-TO-WORK SURVEY

With the co-operation of the British Airports Authority and the airport employers, a survey was conducted in October 1970 to assess the potential demand for a personalized bus service and to indicate the areas in which such a service would be most likely to succeed. Questionnaires were distributed to almost all people working at the Airport on a sample day, either through their employers or by interception of arrivals at car parks or bus stops. Forty-four per cent of the 28 600 questionnaires distributed were completed and returned. These provided detailed information on home address, starting- and finishing-times at work, and the cost, walking, waiting and in-vehicle times of the work journey on the survey day. The questionnaire also outlined a possible personalized service, i.e. 'staff would be picked up near home and taken direct to work; after work they would be collected and taken straight home; the buses would run regularly every day; every passenger would be guaranteed a seat, which would have been booked and paid for in advance'. Respondents were asked to indicate the likelihood of their using a service at a fare level similar to existing public transport. They were also asked to suggest an appropriate weekly fare for their own particular journey, and to state if they would prefer to pay weekly or monthly at a discount.

The Airport comprises four main working areas, viz.

North Side

Maintenance Area

Cargo Terminal

Central Terminal Area

The North Side is a very dispersed employment area, so that the numbers of people who could conveniently be served by one stop were comparatively small. The possibility of operating a personalized service to this area was therefore not investigated at this stage. In the case of the Maintenance Area, the overwhelming majority of workers were employed by only two companies, B.O.A.C. and B.E.A. (now called British Airways). In addition, about 1000 employees of Champion Sparking Plug Co. Ltd., adjacent to the Maintenance Area, could also benefit from the proposed service. The survey there was thus limited to employees of B.O.A.C., B.E.A., and Champion Sparking Plug Co. Ltd. The exclusion of the small numbers of other Maintenance Area staff is not thought to have biased the results in any significant way. In the Cargo Terminal and the Central Terminal Area, all workers were included in the survey.

3.2.3. ANALYSIS OF THE JOURNEY-TO-WORK SURVEY

Summary of Analysis: The first stage identified possible routes for a personalized bus by studying the distribution of respondents by residential area, using London Traffic Survey (LTS) Traffic Zones as a base. It was clear from this that the most promising routes were from Stanwell and Ashford, south of the Airport, to the Maintenance Area. Not only was there a concentration of potential users in these areas but the existing public transport facilities, mainly the 203 bus, were known to provide a poor level of service to this part of the Airport.

Possible routes for the personalized service were drawn up in detail by the LT Traffic Superintendent. These required two vehicles; one would serve Ashford, the other Stanwell, both making at least two journeys each day. For the purpose of analysing the survey data it was necessary to postulate routes with specific timings and stopping places. However, it was envisaged that these arrangements should be flexible so that the precise routing, stopping places, etc. could be adjusted to meet the needs of the actual passengers reserving places in any particular week.

The second stage of the analysis developed a model of travel behaviour which could be used to predict demand for the postulated personalized bus routes. The model assumed that the choice of mode within each zone would be a function of the perceived generalized costs of the available modes. The generalized costs for a zone included a valuation of times spent walking, waiting, and travelling as well as direct fare and car running costs. A modal choice equation was thus calibrated for the existing public/private transport choice. This equation was used to forecast the probability that a car user, faced with the choice of travelling to work by car or by personalized bus, would use the new service. The equation was also amended to include the probability that an existing bus user would switch to the experimental service.

Calibration of the Modal Choice Equation: A full discussion of the method used can be found in *LT Technical Note TN 11*—Modal Split Prediction for Heathrow Airport Staff (1). The description below is based upon *LT Operational Research Report R173* (2).

It was assumed that when there was a choice of more than one mode for the journey to work from a particular zone, the number of people who made the trip by each mode would be a function of the perceived 'generalized cost' of each mode. Generalized costs included a valuation of the times spent walking, waiting and travelling as well as direct costs such as fare and petrol costs. The generalized cost for a zone was based on the average of each element of the individual's generalized cost. The zones were sufficiently small for there to be little skewing of the distribution of individual parameters on any single day.

The generalized cost of public transport for a zone was defined as:
    value of walking time X average walking time for public transport users;
  + value of waiting time X average waiting time for public transport users;
  + value of travel time X average travel time for public transport users;
  + average fare for public transport users.

Walking and waiting times and the fare for each individual would be drawn from the questionnaires. Travel times were calculated by subtracting walking and waiting times from the total time spent travelling to work (i.e. difference between time arrived at work and time left home).

The generalized cost of private transport for a zone was defined as:
    value of walking time X average walking time for private transport users;
  + value of waiting time X average waiting time for private transport users;

+ value of travel time × average travel time for private transport users;

+ average 'cost' for private transport users.

For an individual who used the 'shuttle bus' between the North Side car parks and the Central Terminal Area, the time taken to get from the car park to the workplace was assumed to be split between travelling on the shuttle bus and waiting for it. Walking time was assumed to be zero. (The decision to include all the remaining time in waiting time did not affect the model as walking and waiting times were given the same value.) The average journey time for the shuttle bus was taken as six minutes. For an individual who did not use the shuttle bus, walking time was derived from the questionnaires, while waiting time was set at zero.

Private transport users were asked about the perceived costs of travelling to work and also about mileage. Research workers have usually been uncertain how respondents interpreted this question and whether or not they gave average or marginal values. It was therefore decided that costs for car users would be calculated in two ways, (i) using the cost given on the questionnaire and (ii) using mileage multiplied by a car mileage cost varying from 1¼p to 2½p.

The model postulated that the proportion of people travelling by public transport was a function of the generalized cost of public transport relative to that of private transport. Several alternative models were tested to find the equation with the 'best-fit'. Two theories of individual choice were tested, one based on 'travel by bus costs x times as much as travel by car' (i.e. ratio form) the other on 'travel by bus costs y pence more than travel by car' (i.e. difference form). In addition, two forms of relationship between the proportion of individuals using public transport and relative generalized costs were tested, a linear one and an exponential one. As stated, the generalized costs for car users were calculated using perceived and calculated monetary costs of travel. Let Z be the proportion of people using public transport, GB be the generalized cost of public transport, and GC the generalized cost of private transport. The equations tested were:

$$\text{(i) } Z = a + b\,(GB - GC)$$
$$\text{(ii) } Z = c + d\,(GB/GC)$$
$$\text{(iii) } \log_e\,(Z/(1 - Z)) = e + f\,(GB - GC)$$
$$\text{(iv) } \log_e\,(Z/(1 - Z)) = g + h\,(GB/GC)$$

with items a to h as constants. These equations were fitted to the aggregated zonal data using regression analysis.

The data required for each zone and destination pair were the proportion of respondents using public transport, the average walking, waiting and in-vehicle times, and the average cost for each mode. The only zone-destination pairs included in the analysis were those with (i) at least 30 respondents, (ii) at least 10 private transport users, and (iii) at least 8 public transport users. Users of other modes were not included in either the analysis or the forecasts. In total 27 origin-destination zone pairs were included in the regression analysis.

The Department of the Environment recommends that for fore-casting purposes the value of in-vehicle time for the journey to work should be 25 per cent of gross personal income and that walking and waiting times should be valued at twice in-vehicle time. The New Earnings Survey, 1970, published in the *Department of Employment Gazette* January 1971, provided figures of average gross earnings for full-time men and women in Greater London. From this a value of in-vehicle time of 17·82p per hour was arrived at and used to calculate generalized costs.

It was found that an exponential form of equation fitted the data better than a linear one and that the stated cost of private transport, rather than a cost computed from mileage times cost per mile, explained travel choice more satisfactorily (whatever value of car mileage cost was assumed). In all cases, relative generalized costs were better expressed in a ratio form: respondents seemed to base their choice on 'The bus costs x times as much as the car'. The 'best-fit' equation for the public/private transport choice was thus

$$\log_e(Z/(1 - Z)) = 1·24 - 0·849\ (GB/GC)$$

i.e. $$Z = \frac{\text{Exp }(1·24 - 0·849\ (GB/GC))}{1 + \text{Exp }(1·24 - 0·849\ (GB/GC))}$$

where Z is the probability of choice of bus rather than car.

It was found that two individual zones covering Ashford and Stanwell had observed values of Z (Z'say) for the car-bus choice that were rather different from those predicted by inserting the values of GB and GC for those zones in the above equation. It was decided, in estimating the new Z_1 for the personalized bus service/private car choice, not to derive the absolute value of Z_1 from a straight application of the above formula, but to calculate the change in Z obtained by substituting the personalized bus costs for the existing public transport costs, and to apply this change in Z to the observed value of $Z'$for these zones.

Forecast use of a Personalized Bus Service: The above equation can be used to predict the probability that a car user, faced with a choice of travelling to work by car or by personalized bus, will use the new service. This is done by inserting the ratio of generalized costs by personalized bus (denoted *GP* in the following discussion) and by car and re-calculating Z_1. For those who already travelled by public transport the equation was amended so that if the generalized cost of existing public transport and the personalized bus were the same (i.e. $GP/GB = 1$) there would be a 50 per cent probability of using the new service. The forecasting equation showing the split between the personalized bus existing public transport was:

$$\log_e (Z_2/(1 - Z_2)) = 0{\cdot}849 - 0{\cdot}849 \ (GP/GB)$$

i.e. $Z_2 = \dfrac{\text{Exp}(0{\cdot}849 - 0{\cdot}849 \ (GP/GB))}{1 + \text{Exp} \ (0{\cdot}849 - 0{\cdot}849 \ (GP/GB))}$

where Z_2 is the probability of choosing the personalized bus rather than existing public transport.

Only respondents in Ashford and Stanwell were included in the forecasting stage of the analysis. To assist the calculation of travelling times, respondents were grouped into 'picking-up points'. This did not mean that these points were adhered to when the service was introduced, since the routes were flexible so that they could be varied to meet individual passengers. Respondents were assigned to the point nearest their home. In most cases the walking time from home to the picking-up point was taken as one minute, although this was increased by a penalty time (usually a further minute) for those respondents who did not live close to the proposed routes.

A timetable was then proposed for each route and the number of respondents who might use the service was calculated from their stated present time of arrival at work and the scheduled arrival times of personalized buses at Heathrow Airport. For brevity, these respondents will be called 'allocated' respondents. 'Allocation' was on the basis that a respondent would use the bus which arrived at the Airport just before his stated time of arrival. If the first bus on a route arrived at the Airport after a respondent's stated time of arrival, then the respondent was not 'allocated' to any journey. Similarly, if the last bus arrived more than fifteen minutes earlier than the respondent's usual arrival time for work, the proposed service was assumed to be unsuitable. It was assumed that the walking time from Hatton Cross, where the bus planned to stop, to the Maintenance Area was two minutes. Respondents who worked in the Central Terminal Area or the Cargo Terminal were excluded from the analysis by setting high walking times from Hatton Cross i.e. 99 minutes. The number of respondents who might use the service on each route was then calculated and this procedure was repeated for several different timetables to find the times of arrival for each service that suited most respondents. Later stages of the analysis then used this timetable.

The stated finishing-times for 'allocated' respondents were then examined to decide the most suitable times for the return journey in the evenings. Respondents whose finishing-times were not covered by the proposed timetables were excluded from further analysis i.e. they were treated as non-users of the service.

To summarize, the only respondents for whom a probability of using the service was calculated were those whose stated times of arrival at and finishing work were compatible with the proposed timetables. A computer programme was written to calculate the probability of each of these respondents using the personalized bus service, using either the personalized bus/car equation or the personalized bus/other public transport equation, as appropriate. The following data were used in this programme:

    (i) time of arrival of first bus at Hatton Cross
    (ii) number of journeys the bus makes
    (iii) number of picking-up points on the route
    (iv) walking time from Hatton Cross to the Maintenance Area, the Cargo Terminal, and the Central Terminal Area
    (v) interval between journeys
and for each picking-up point
    (vi) point number
    (vii) cost of a single journey if payment was made weekly
    (viii) cost of a single journey if payment was made monthly
    (ix) walking time to picking-up point
    (x) journey time from picking-up point to Hatton Cross
    (xi) penalty time for respondent who does not live near the route.
It was also necessary to use the observed proportion of respondents in a particular zone using public transport (for public/private transport choices).

The probability of using the service was calculated by comparing the costs and times of travelling by personalized bus and by the mode presently used. The fare a respondent paid on the new service was assumed to be dependent on the answer to the questionnaire; if a respondent stated that he would pay monthly in advance, then he was charged accordingly. A discount of approximately 20 per cent was assumed for tickets bought monthly. The following information was available from the programme for each potential user:

    (i) present mode of travel (car or public transport)
    (ii) place of work (Maintenance Area, Cargo Terminal, or Central Terminal Area)
    (iii) journey to which the respondent was 'allocated'
    (iv) stated use of service
    (v) forecast use of service

(vi) stated time of leaving work

(vii) stated time of arrival at work.

In addition, a summary of the forecast number of respondents using the service from each picking-up point was given. The programme was run for several sets of fares on the personalized bus in order to estimate how demand might respond to changes in fares and to suggest the most appropriate fare level.

The forecast number of passengers was calculated for each journey and for each fare structure. For purposes of comparison an estimate of the number of passengers on each journey was prepared, based on the declared use of the service. Only those respondents who answered 'definitely' or 'probably' (given a weight of two-thirds) or 'possibly' (given a weight of one-third) were included in this estimate. It was thought that those respondents who declared they would 'probably not' or 'definitely not' use the service may have had personal reasons for using private transport which made the personalized service inappropriate.

All the results provided by the computer programme were grossed up to include those respondents who were not included in the analysis because of missing or inconsistent data on the returned questionnaires.

At weekly fares of £1 from Stanwell and £1·50 from Ashford, the forecast numbers of personalized bus users were 97 from Stanwell and 52 from Ashford. These were thought at the time to be pessimistic estimates since they had been obtained without grossing up the sample of respondents to the total population, i.e. the 56 per cent of Airport employees who failed to complete the questionnaires were excluded from the analysis. (It was assumed that serious potential users were more likely to have returned a completed questionnaire.) Furthermore, the results obtained from the modal choice analysis appeared to be pessimistic compared to the direct question on whether respondents would or would not use the service. The effect on modal choice at these fares was expected to be:

	% Car	% Personalized bus	% Other bus
Stanwell			
Before	82	–	18
After	62	23	15
Ashford			
Before	86	–	14
After	68	21	11

Receipts were expected to fall short of direct operating costs by about 14 per cent. The Department of the Environment agreed to regard the service as a Bus Demonstration Project and to bear all marketing and survey costs and any operating deficit incurred in the first six months' operation up to a pre-determined level. London Transport would finance any further operating deficit.

3.3. THE SERVICE

The service was introduced on 15 January 1973 under the name 'Skyport Special'. With minor modifications, the service was similar to that postulated in the modal choice analysis mentioned above. The vehicle used was an 11 metre long coach hired from a local operator. A special office was set up at the Airport to take bookings and control the running

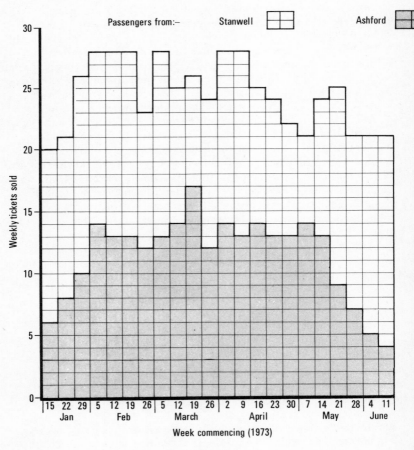

Fig. 3.1

of the service. The experiment was publicized by the distribution of leaflets to houses in Ashford and Stanwell and through the major employers at the Airport. The numbers of passengers carried are shown in Figure 3.1. Within four weeks these built up to about 40 (compared to a forecast of 149) which was thereafter maintained until the onset of the holiday season, from which point it declined until the service was withdrawn on 15 June 1973. Total receipts for the 22 weeks' operation covered less than 20 per cent of direct operating costs.

3.4. THE 'POST MORTEM'

3.4.1. METHODS EMPLOYED

The original intention was to carry out a quantitative survey after the service had been in operation for three months. Short interception interviews were therefore carried out in the Maintenance Area at Heathrow to attempt to collect the names and addresses of all employees who either used the service or could possibly have done so by virtue of their home location and time of arrival at work.

The response to these interviews was disappointing both in terms of the total number of suitable staff intercepted and the proportion of those who were agreeable to a subsequent interview at home. In all only 209 relevant names were obtained and it was decided not to proceed with a quantifiable survey suitable for the L.T. model, but to carry out an in-depth attitudinal survey with a limited number of respondents.

The attitudinal survey yielded 105 successful interviews as follows:

35* Skyport Special users
9* Skyport Special ex-users (now using various modes)
5* Scheduled bus users
33 Car drivers
23 Car passengers
___
105

(* these represent the maximum possible sample size in these categories).

The questionnaire consisted of predetermined questions asked by the interviewer but responses were recorded verbatim, without precoded responses, and were analysed later. The sample sizes in each category were too small for the results to be statistically valid but they are more than adequate, particularly in the cases of the Skyport Special users and car drivers and passengers, to allow a comprehensive assessment of the prevailing attitudes of the population sampled.

The attitudes of those interviewed were naturally affected by their attitudes to the other modes available for their use. Users of the service were therefore also asked about their previous mode and non-users about their current mode of transport to work. General questions about

Heathrow Airport as a work location and its transport facilities were also included in the questionnaire. A full report of this study is contained in Freeman Fox and Associates' Report 'The Skyport Special, Attitudinal Surveys to monitor the Heathrow Personalized Bus Service' (3).

3.4.2. ATTITUDES OF USERS OF THE SERVICE

Over 60 per cent of the actual users of the Skyport Special had previously travelled to work on the 203 scheduled bus. The majority were women and nearly all were clerical—secretarial workers or in skilled manual occupations.

All users of the Skyport Special were extremely enthusiastic about the service, and it clearly provided an attractive means of getting to work. Very few respondents thought that the service could be improved, and the majority would have been willing to pay more for it, indicating satisfaction with the service.

The respondents who had previously been scheduled bus passengers were particularly positive in their reactions; they all found the service reliable, quick, and convenient. They liked especially the door-to-door nature of the service. They were nearly all comparing the service with the scheduled bus no. 203 which is generally thought to provide an extremely low level of service over the same general route. None found that the journey by Skyport Special took longer than by the more directly routed scheduled bus; 90 per cent made time savings over their previous journey, saving an average of about half an hour.

The smaller number of respondents who had changed from going to work by car were also enthusiastic about the Skyport Special, finding it advantageous not to have to drive on what they considered to be congested roads and find a parking space. These users were comparing it with the previous car journey when they voiced some complaint at the circuitous routing and the need to leave work at the same time every day. However, these views were only mildly expressed compared to some later respondents who did not use the service. It is interesting that the majority of the ex-car users benefited neither in terms of cost nor time by using the Skyport Special (i.e. the modal choice model would not have assigned them to the Skyport Special), yet all were enthusiastic about the service. This suggests that some people do indeed find driving on congested roads, or finding a parking place, a deterrent to commuting by car and that a public transport service of high quality does provide an attractive alternative.

However, it was evident that none of the users worked overtime on a regular basis and that they kept very regular working hours. The pattern of the age, sex, and socio-economic structure of the service users implied that they probably had a well-settled home life to go back to in the evening. They were thus an ideal clientele for the pre-booked fixed time services provided and contrasted in this respect with the majority of the airport workers from Ashford and Stanwell.

3.4.3. REASONS FOR CEASING TO USE THE SERVICE

Nine interviews took place with workers who had tried the Skyport Special but had stopped using it. Although the number was too small to permit any firm conclusions, it was interesting to note that there was no criticism of the basic service. For particular reasons, which were associated with the necessarily limited experimental service or for independent personal reasons, these people had abandoned the Skyport Special. However, four had done so with regret, and had found no major personal disadvantages in using the service.

These latter respondents had changed to using cars, two as drivers and two as passengers. Those returning to driving had come under pressure from their respective spouses, one of whom had lost a lift to the railway station through her husband using the Skyport Special, while the other did not drive himself and resented his wife not utilising a car for which he helped to pay. The two who returned to being passengers had been only temporary users of the Skyport Special while their regular lift-giving driver was away. Both would have preferred to have continued using the service but did not wish to 'let their driver down', presumably by withdrawing their financial contributions to the cost of the car journey.

The remaining five ex-users of the service had valid but varied reasons for returning to the previous mode—three driving, one a car passenger, and one a cyclist. Because of the routing pattern the cyclist suffered a severe time disbenefit when using the Skyport Special; while prepared to suffer a cost disbenefit to achieve comfort, he was not prepared to accept a longer journey time. This was exacerbated by some recent evenings of overtime when he had been unable to utilize his pre-paid ticket and had to find an alternative way home. The three respondents returning to driving their cars had particular problems in dovetailing working hours to the scheduled running times of the Skyport Special. This was due to the limited provisions of the experimental service. The only abandonment on cost grounds came from a young girl in difficult financial circumstances who accepted the offer of a free lift from a friend, despite problems when one or other of them worked overtime.

3.4.4. ATTITUDES OF WORKERS NOT USING THE SERVICE

Shared cars: London's Heathrow Airport has specific features as an employment centre which encourage the use of the car for travelling to work. The provision of large-scale car parks in the Maintenance Area and the poor public transport services to that part of the Airport make the car, if available, a highly attractive mode for these workers for whom the

Skyport Special was provided.

Allied to this is the exceptionally high level of the giving or accepting of lifts on a regular basis. This well-established pattern encompasses both family groups and sets of work colleagues. The attraction of Heathrow as a local employment centre is illustrated by the number of instances of spouses or siblings also working at the Airport. In some cases the sharing of a car by two or more family members reinforces the modal choice; alternatively the availability of a lift may be the deciding factor in a wife choosing to go out to work at all. In either event the sharing of the car to work is more part of the life style than purely a means of transport.

Given the high level of car usage and a public transport facility which has built up a reputation for unreliability, it is not surprising that a place as a passenger is anxiously sought by advertisement on office notice-boards and is jealously guarded once obtained .

However, car passengers and drivers carrying them have perforce to keep regular hours and in this respect this category of respondents had a pattern of travelling which was closer to the Skyport Special service than that of other non-users. Few worked overtime or late and most were attuned to having to be ready at a specific time. The Skyport Special was therefore perceived as having few disadvantages, although conversely neither did it offer any great advantages over their current mode.

About two-thirds of the car passengers had lifts from members of their family; their journeys were easy, reliable, quick, and, in most cases, cost nothing. Consideration had been given to using the Skyport Special only in cases where there were problems of synchronizing times with the driver or if the lift was in one direction only. However, on relinquishing the family lift through change of circumstances, they would have been very likely to use the Skyport Special in preference to seeking a lift from outside the household.

The remaining car passengers obtained lifts from fellow workers and in most cases admitted that they contributed financially to the cost of the journey. They continued to travel as car passengers from habit and because they were unwilling to cancel an existing arrangement. Several of them mentioned spontaneously that they would change to the Skyport Special as and when their lifts ceased, for example at the beginning of the holiday season or when their driver changed his work pattern.

Cost was not a major factor for most car passengers despite the fact that many of them paid nothing for their lift. Perhaps for this latter reason they had no yardstick by which to measure the cost of the Skyport Special but perceived it as attractive and were willing to pay the asking price.

The car passengers were more knowledgeable about the details of the service as published by LT than any other group of non-users. It was a common theme amongst most non-users of the service that the schedule did not fit their travelling times, although this was frequently not true. The explanation may lie in a subconscious reluctance to believe that the service could be appropriate for the respondent's need or the reflection of an unexpressed opinion that the service was not sufficiently flexible to cater for the occasional few minutes of working late or any delay in leaving the workplace.

It rapidly became apparent that inflexible departure times were the most crucial factor affecting the decision to use the Skyport Special, followed closely by the need to book (and pay for) 10 journeys per week.

Bus users: The small number of respondents who continued using the 203 bus after the introduction of the Skyport Special service did so either because the service did not match their working hours closely enough (although considered possible users by the Study Team) or because of regular lifts in the homeward direction on several days of the week. Again, they did not appear to be excessively cost-conscious but were very conscious of how wasteful it would be to pay for the Skyport Special (at a premium over the scheduled bus fare) and then have to incur fares on the occasions when they were unable to use the service.

Car drivers: The lack of flexibility was the most strongly voiced comment by almost all the car drivers interviewed. The flexibility and convenience of the car was seen to be its greatest advantage and was offered as the major reason for choosing it by a large majority of drivers.

Overtime and working late, while not generally very frequent, is an intrinsic part of many of the jobs at the Airport. Most of the respondents considered that such occasions could and did arise without prior warning and that having a car available for the homeward journey was a virtual necessity. However, when the problem of working overtime was raised in connection with the use of the Skyport Special the concern was not over the problem of getting home but of having to pay in advance for a journey that might not subsequently be made.

All drivers knew that the service has to be booked for a week at a time and this was a consistent cause for complaint. Several alternative suggestions were made for the booking system, some of which could perhaps have been tried out had the experiment been continued.

However, while the pre-booked system was the greatest cause for complaint, not even the most radical alteration could have persuaded more than a small proportion of the car drivers to change to using the Skyport Special.

It appeared that about half of them used their cars during the day, either occasionally or regularly; a number either had regular evening commitments to which they went straight from work or they occasionally stopped on the way home. Thus the car was utilized in a variety of ways between leaving home in the morning and returning there at night. This included lunchtime or evening shopping trips, trips home or to a pub for lunch, social and recreational trips in the evening, and taking other family members to school or the station. Some of these trips are indicative of the lack at the Airport of the facilities found in most town centres, others are inspired by the poor nature of public transport services in the area, while some are generated because the car is available. Lack of freedom and independence is a major deterrent to the committed car user considering an alternative mode, but in this case there is a high level of supplementary journey making that would have been genuinely inhibited by use of the Skyport Special.

Previous studies of industrial workers have shown that the manual worker is generally much more cost and time conscious than other social groups, so that generalized cost models mirror his behaviour quite closely. White-collar workers, on the other hand, seem to have enlarged sets of criteria for making modal choice decisions. In the present study the respondents were found to be aware of costs, as might have been expected, but both manual and non-manual workers were also aware, and highly appreciative, of other attributes of the car mode which are not generally included in modal choice models. The geographical position of the Airport and the general acceptance there, by workers and policy makers alike, that the car is an appropriate mode for the work journey and should be catered for by providing ample parking space and good access roads, combine to reinforce the natural inclination of the car owners to drive to work. It is most often the availability of the car and the absence of any major disadvantages to its use which cause it to be used in preference to a public transport system of whatever standard. There is little perceived deterrent to the use of the car to go to work at the Airport and disadvantages such as wear and tear on the car or journey cost, although perceived by some respondents, were willingly traded for the much greater advantages presented by car use. With a poor existing public transport system a commuter to the Airport does not need to be a 'committed car user' to have become a regular car user with little perception of the relatively minor disadvantages to himself involved.

The Skyport Special was regarded as intrinsically another form of public transport, with its slightly different features rendering it more or less attractive depending upon the viewpoint of the user. It was not generally perceived as an alternative mode for car drivers, who looked upon it as an attempt to provide a better-quality service for existing public transport users.

3.5. RERUN OF FORECASTING MODEL

In view of the large discrepancy between the estimated (149) and actual (40) number of users of the service, further work was carried out in an attempt to identify some of the major factors contributing to the over-estimation by the behavioural model. This is described in detail in *LT Operational Research Memorandum M276* (4).

The actual service provided was slightly different from that assumed in the forecasting model. The two routes specified were actually combined so that the service ran from Ashford via Stanwell to the Maintenance Area, with the result that the average journey time for Ashford passengers increased. On the other hand, passengers had a choice of four rather than three buses in the morning. Monthly tickets at a discount were not introduced so that the average fare also increased.

The forecast traffic had been based primarily on the characteristics of the journey to work in the morning. No attempt was made to study the sensitivity of the estimates to the scheduled departure times in the evening.

The forecasting model was rerun to reflect the actual service offered by the Skyport Special and to test its sensitivity to other assumptions. In the initial stages one amendment at a time was made to identify any major influences. Firstly, the actual scheduled running time and cost of the Skyport Special were included. Each of these appeared to have little effect on the forecast. Altering the arrival times of the morning buses reduced the expected usage by 3 per cent (to 145 passengers); there was a similar reduction when the journey time for Ashford passengers was increased. When the possibility of buying monthly tickets at a discount was removed the estimate fell by 2 per cent (to 146).

The sensitivity of the forecast to restrictions on the finishing-time of potential passengers was also tested. As the departure times of the Skyport Special were between 16.30 and 17.15, only those employees who finished work between 16.20 and 17.15 were included in the forecast; this factor reduced the expected number of passengers by 37 per cent (to 94).

Further attempts were then made to include the perceived flexibility of the car. The methods were necessarily rather crude because of the intangible nature of the car's benefits. In order to reflect the inconvenience of paying for ten journeys each week it was assumed that each passenger would miss one journey each week so that the price per journey was increased. This reduced the forecast usage by 4 per cent (to 143).

In the original model no account was taken of the extra waiting time before starting work that might be incurred by using the Skyport Special. Two ways of introducing this were tried, both of which had an enormous impact on the forecast. Firstly, any passenger who would arrive by the Skyport Special more than (a) 10 minutes, (b) 15 minutes, and (c) 30 minutes earlier than he did by his present mode was excluded from the forecast.

This reduced the estimate of usage by 43 per cent, 30 per cent, and 4 per cent to 85, 104, and 143 passengers respectively. The second, slightly less arbitrary, method was to assume that car users at present arrived at work at the exact time of starting, that the present bus users incurred no additional wait, and that this latter waiting time should be included in the valuation of the journey to work. These assumptions reduced the estimated usage by 24 per cent to 113 passengers.

Finally, the forecasting model was rerun including all the above assumptions and restrictions at the same time:

(i) amending times of arrival of the Skyport Special

(ii) increasing journey times from Ashford

(iii) removing discounts and assuming fewer journeys per week

(iv) restricting the finishing-times of possible passengers

and

(v) imposing a maximum waiting time at work of 30 minutes and including the wait in the 'generalized cost' of the journey.

This reduced the estimated usage of the Skyport Special by 70 per cent to 45. The result of this decrease was to remove most of the discrepency between the model forecast and actual patronage, which had stabilized at around 40 passengers.

One of the important factors not included in the revised forecasting model was the usefulness of the car for making extra journeys during the day, e.g. to the shops. Approximately half the car users interviewed in the attitudinal survey claimed to use their car during the day and this was probably an important factor in their choice.

3.6. CONCLUSIONS

3.6.1. SPECIFIC CONCLUSIONS RELATING TO THE SKYPORT SPECIAL

Specific conclusions drawn from the 'Post Mortem' covered aspects of marketing, fares and ticketing, flexibility of operations and booking, choice of vehicle, area of operation, and choice of appropriate market segments.

Marketing: Whilst the marketing prior to and during the launch of the service was adequate and effective, it would have been better to have had a higher level of continuing publicity. Since the service represented a radical departure from that offered by the established bus service, it would have been valuable once the levels of usage had begun to stabilize to launch a secondary campaign utilizing as an additional marketing feature the high level of appreciation expressed by those people who had chosen to use the service.

Fare levels and ticketing system: There was little evidence to suggest that the fare levels were an active deterrent to using the service and it was thought that the fares set were at an appropriate level and that the ridership would not have increased significantly if they had been reduced.

There was nevertheless some evidence which suggested that some non-users were deterred from using the service because they resented the idea of paying for odd missed journeys. To offset this criticism it would have been beneficial to try various alternative systems of payment had the experiment carried on for a longer period of time. The most practical method appears to be a period ticket at a discount: the discount would be perceived to represent a certain number of free trips which could either be taken as a bonus or missed without penalty.

Flexibility of operations and bookings: One of the principal criticisms of the service made by non-users was its lack of flexibility and the need to book in advance. The criticism was particularly pertinent to the evening trips when many of the respondents felt the need to have some flexibility about when they could leave work.

The *modus operandi* used was somewhere between that used for a scheduled fixed route service and a totally flexible Dial-a-Ride service, and as such ran the danger of falling between two market sectors.

It would have been possible to counter much criticism, particularly from car users, by making it more responsive to demand. In order to accomplish this objective it would have been necessary to have a facility whereby the users could ring up the service-controller during the afternoon in order to book a trip for that evening and the following morning.

It is probable that a majority of users would have used the same services almost every day so that the administrative problems created by the additional flexibility of the system would not have been unmanageable. The options outlined above would have made the service more attractive to potential users. Greater demand responsiveness would nevertheless have put up administrative costs.

Choice of vehicle: The vehicles used were single-decker, 11 metre long coaches, hired from a local coach operator. The marketability and attraction of the system could have been improved by the use of smaller buses which could more easily have penetrated residential areas. While there are disadvantages associated with the operating costs of such vehicles, they have greater passenger appeal, facilitating the presentation of a distinct 'brand image', and could have better catered for the levels of patronage which finally emerged.

Area of operation and market segmentation: The low level of usage from the areas served is largely attributable to the attitudes associated with private modes. Although the residential pattern of workers in the Central

Area is more diffuse than that of the Maintenance Area, the policy of parking restraint in the Central Area, leading to the use of a shuttle bus to distant car parks, should make it a more fruitful area for an experimental public transport service of this kind.

Some potential users were also not catered for at all because the experimental service only operated during certain morning and evening hours. Shift workers are normally ruled out when such services are considered, although it may have been worth revising this attitude in the case of Heathrow, where shift workers have their work journeys concentrated into distinct time periods and generally adhere closely to the scheduled starting- and finishing-time. Running services throughout the day would also have made the economics of the operation more favourable, while complementing the more skeletal off-peak scheduled bus services.

3.6.2. GENERAL CONCLUSIONS

A number of general conclusions can be drawn about the operation of this type of service and the methods of passenger forecasting used. These relate to the timing of the experiment, the experimental nature of the service, the choice of scenario, and the approach to the modal choice modelling.

Length of experiment: It was unfortunate that the very low levels of utilization caused the service to be withdrawn at such an early stage. Adjustments to the service to overcome some of the perceived disadvantages would have had some effect on ridership; these alterations would necessarily have had to allow time for the new patterns of use to stabilize.
Recommendation by word of mouth played some part in increasing patronage but it takes time for dissemination, acceptance and actual change to take place. It also seems likely that a small proportion of car passengers would have changed over to the service over a period of some months.

It seems, therefore, that an experiment of this kind must be sustained over a period of several months and must incorporate some method for adjusting the detail of the operations to meet customer requirements.

The experimental nature of the service: It is difficult to generalize about the attractiveness of a full-scale, all-day service on the basis of such a limited experiment. There was little adverse comment in the attitudinal survey to the experimental nature of the service; however, some potential users intercepted at the start of the 'Post Mortem' refused to be interviewed on the grounds that an experimental service of limited duration was of no interest to them, even to the extent of granting an interview.

Choice of scenario: It is clear that the success of an innovatory public transport system cannot be guaranteed, however high the level of service or whatever the length of the experiment, in a scenario which does not restrain use of the private car. This is particularly true when that scenario has been in existence for several years so that habits have formed which cannot easily be changed. This suggests that the optimal place for such services is in areas of recent change, either of new development or of changes in parking or road congestion.

Approach to the predictive modelling: In the context of this anthology of experimental evidence the most important general point of interest is the large discrepency between the forecasts and the actual patronage. The model assumed a value of time of 17·82p per hour for in-vehicle time and double that for walking and waiting times in the calculation of generalized costs (the ratio of which was used to forecast the modal split). The model was then calibrated on the basis of the journey to work. Behaviourally, it has been observed that for many commuters, particularly manual workers, it is delays in the journey *from* work which are more irritating. Although the journey to work has to facilitate a prompt arrival (and thus both reliability and journey time are important), delays in the journey home are perceived as an intrusion into private leisure time imposed by the transport system and arising from the work process. In this sense the homeward journey time is seen as time unwillingly given to the employer and, particularly if the job is tiring, the traveller is more inclined to be irritated by delays at the end of the day. It could therefore be hypothesized that time spent waiting for a bus coming from work is valued more highly than when going to work, and that thresholds of delay exist beyond which a potential passenger is not prepared to go. The major sensitivity identified from the attitudinal study was the relationship between bus departure-times and work finishing-times, and it is interesting to note that when the model was rerun to exclude workers with a wait of more than 10 minutes at the end of the day, the forecast ridership fell by 37 per cent (55 passengers), a greater effect than all the other alterations combined (a 10 minute delay at the beginning of the day caused ridership to fall by 30 per cent).

The model used was a straightforward generalized cost model calibrated from observed data. The rerun of the model which eventually produced results close to the achieved levels of ridership, incorporated in a fairly *ad hoc* way those items which had been found to be important from the attitudinal research. Some changes in model output were obtained by adding extra weighting factors, while others arose from using a threshold approach whereby certain people were excluded because one or more of their parameters lay outside a prescribed range.

From this exercise it is possible to envisage a situation in which such factors are incorporated at the earlier, forecasting stage of modelling.

Cases where predictions are checked and followed up by reruns of the model are rare. However, a significant amount of work has been carried out in relation to factors affecting modal choice, apart from the basic elements of time and cost. It should therefore be possible to draw up broad guidelines for the sophistication of modal choice models. The relative importance of different elements of generalized cost can readily be investigated. A more difficult and subtle exercise is the establishment of thresholds, or cut-off points. Existing models are based on smooth curves; there is a large amount of evidence to suggest that a 'stepped' curve, with distinct break points, would be more appropriate, although the thresholds are likely to differ between journey types and lengths. The work described in this chapter indicates one direction in which further research into modal choice modelling could proceed.

REFERENCES

1 London Transport, Modal Split Prediction for Heathrow Airport Staff, *LT Technical Note TN11*, mimeographed 1971.

2 London Transport, Personalized Bus Services for Heathrow Airport Staff, *LT Operational Research Report R173*, mimeographed 1971.

3 Freeman Fox and Associates, The Skyport Special: Attitudinal surveys to monitor the Heathrow Personalized Bus Service, unpublished report 1973.

4 London Transport, Heathrow Personalized Bus Service: A comparison of predicted and actual patronage, *LT Operational Research Memorandum M276*, mimigraphed 1973.

5 Heraty, M. J., and Day, D. J., Experience with a Personalized Bus Service for Heathrow Airport Staff, *PTRC Summer Annual Meeting*, Warwick, 1974.

4

Valuation of Commuter Travel Time Savings: An Alternative Procedure

by D. A. HENSHER

4.1. INTRODUCTION*

According to the neo-classical rationale, in the absence of any market imperfections individuals will allocate their time between activities in such a way that the value of time at the margin for all activities, including travel, will be equal to the marginal wage rate. Thus any reduction in the time spent in any activity, should be valued at the wage rate regardless of whether the time savings were achieved during working hours. Since the marginal wage rate may be defined as net earnings received for the marginal unit worked, it can be referred to as the average hourly earnings received. Hence the rational individual will 'buy' time by using quicker transport at the same rate that he sells time to his employer. This assumption is based on at least two restrictive assumptions:

1. The individual is able to make marginal adjustments between work and non-work activities.
2. The choice he makes is only related to money (the marginal net wage) and time (unit of leisure or non-work), thus ignoring any non-monetary aspect of work such as the involvement, experience, and enjoyment derived from the work itself.

This approach implicitly assumes that the marginal utility of work is zero, so that the marginal wage compensates the worker only for the

* The assistance given by the Commonwealth Bureau of Roads in making it possible to develop and test the arguments of this chapter is gratefully acknowledged, although the opinions expressed do not necessarily represent the views of the Bureau of Roads. Appreciation is also expressed to many persons for useful comments and discussion on earlier versions of this chapter (7), in particular to Ian Heggie, Reuben Gronau, Ken Rogers, David Quarmby, Ted Delofski, Paul Mcleod, and Chris Nash.

marginal sacrifice of leisure. This is the Jevons—Robbins doctrine, that the marginal utility of leisure, or the marginal disutility of labour, is zero. Since work involves disutilities (e.g. poor working conditions) and utilities (e.g. the experience gained and the greater potential enjoyment of leisure time because of the psychic satisfaction associated with earning money for use in future leisure time), and the marginal net utility of leisure must be positive (otherwise men would work until the marginal wage rate fell to zero or would continue working round the clock), then the wage rate represents more than just the value of the employee's time. It also represents the net disutilities of labour incurred. Relaxation of these assumptions is usually taken to imply that the marginal valuation of leisure time is less than both the gross and the net marginal wage rate. That is, it implies that people are on average able to work less hours than they would if they could choose, or have a negative marginal utility of labour such that the marginal wage minus the marginal disutility of labour is equal to the marginal utility of leisure.

The individual is assumed to attempt to maximize his perceived utility subject to both money and time constraints, and since there is a fixed amount of resource time (i.e. 24 hours a day), the marginal value of travel time will rise as demand for consumption activities increases. As real incomes increase, individuals will therefore tend to substitute more income for less time, i.e. their propensity to spend money to save time increases more than proportionately with changes in their income.

The neo-classical wage rate argument outlined above presented a simple, convenient technique for valuing travel time and was used until the early 1960s when it was formally and empirically placed in jeopardy (5, 12).

In the belief that the behavioural value of private travel time savings is less than the average wage rate, numerous studies (8) have been undertaken to calibrate models capable of estimating the value of travel time savings, and in particular the value of commuter travel time savings. In these studies the technique adopted has not differed, with a few exceptions (13), but there have been differences in the internal structure, (perceived versus synthetic), external structures (differences versus ratios etc.) and disaggregation (waiting, walking, in-vehicle times etc.) of the trip variables. Many of the studies have applied the techniques previously used in commuter travel studies to other purposes of travel without questioning the validity of generalizing techniques. Nearly all the criticism of existing studies has related to the data and the statistical tools available (discriminant analysis versus logit analysis versus probit analysis etc.), but in general not to the conceptual procedure itself.

Since one of the major economic benefits from many transport investment projects is private time savings (2), it is important that the values used have some rationale. The level of such investment is very sensitive to the value of travel time. For example, in the U.S.A. (18), with a value of travel time savings of zero, the Transportation Resource Allocation Study (TRANS) models indicate that investment should be reduced by about 40 per cent from that indicated by the base value ($3·00 per private vehicle hour and $6·00 per commercial vehicle hour). Doubling the base value results in an increase of about 66 per cent in investment.

This chapter is concerned with the conceptual and empirical development of an alternative procedure to the behavioural (as distinct from resource) valuation of commuter travel time savings. The basis of this alternative method is expressed in two hypotheses related to the underlying binary choice process which leads to the selection of alternative combinations of commodity characteristics and to the choice situation facing an individual. The argument is not so much that the alternative procedure is a definite improvement over the traditional approach (8) as that it is better to consider the acceptability of a series of alternative hypotheses on individual traveller behaviour. It can be shown, however, that there are definite inconsistencies between the generally accepted definition of the behavioural value of travel time savings and the appropriate interpretation that should be given to the empirical values of travel time savings derived from current statistical models. Alternative definitions of the value of travel time savings are proposed and discussed.

4.2. APPROPRIATENESS OF THE 'TRADITIONAL' APPROACH TO TRAVEL TIME VALUATION

The only 'successful' method of valuing private travel time savings has been by comparing the coefficient of travel time with the coefficient of travel cost in a modal choice (2 modes) or route choice (2 routes) framework. From a simple equation of the form

$$P_1 = \frac{e^y}{1 + e^y}, \text{ where } y = \alpha_0 + \alpha_1 (t_1 - t_2) + \alpha_2 (c_1 - c_2) \tag{1}$$

where P_1 = probability of choosing mode (route) 1
  y = choice of mode (route): 1 = mode (route) 1; 0 = mode (route) 2
  e = exponential constant
  t_i = door-to-door travel time by the ith mode (route); $i = 1, 2$
  c_i = door-to-door travel cost by the ith mode (route); $i = 1, 2$

a value of travel time savings has been inferred by noting the changes in the dependent variable resulting from a unit change in either the time or the cost difference. Thus, a unit change in the time difference will be associated with a change of α_1 units in the regressand. The same change in the regressand could be obtained by a change of α_1 / α_2 units of cost. Hence a value of travel time savings may be inferred as α_1 / α_2 from this disaggregate binary choice model.

The derivation of the value of travel time savings from the apparent time–cost trade-off in a mode or route choice model may not be an appropriate approach. Disaggregate binary choice models do not test any specific hypothesis about the value of travel time savings; they merely test the hypothesis that money cost, time, and any other explicit variables are important in determining variations in modal choice. With known values of travel time savings, however, it may be possible to test hypotheses in a somewhat different sense; for example, in explaining choice of mode, how plausible is it that the value of time is an increasing function of income? (3,6) The importance of the money costs associated with mode choice is not necessarily the same thing as the general import-ance of money to the individual: 'the real worth of things to a man is not gauged by the price he pays for them' (16).

Given an hypothesized cost–time saving relationship, where the relationship itself is an important analytical assumption of the model, it may be methodologically and statistically undesirable to specify a model where both cost and time savings are explanatory variables. It might therefore be useful to investigate the reactions of travellers to time savings in other ways, by developing hypotheses in accordance with the definition of the value of travel time savings and then to examine, using the appropriate statistical technique, the extent to which the observed behaviour of all individuals is consistent with the hypotheses (9). It is meaningless to talk about how consistent or inconsistent an individual's behaviour is in relation to the hypotheses being tested. The technique should only be used for the testing of hypotheses, not for their formulation. An hypothesis that has been shown to provide an adequate explanation of behaviour may not be the 'best' model. There is no way of knowing when it is (3).

4.3. ALTERNATIVE APPROACHES

There are three basic ways in which we might investigate the relationship between the 'ingredients' of the value of travel time savings; firstly, by means of the above traditional approach; secondly, by incorporating hypotheses about the value of travel time savings into modal or route choice models (e.g. the value of travel time savings is a function of the individual's personal income); and thirdly, by measuring the value of travel time savings directly.

We have already commented on the first approach. The second method is limited by the need to have a clear idea of any value of time dependence that might exist, and assumes that the impact is the same for all states of the modal variables (6, pp. 259–60). This method might be useful as a

mechanism for separating out the influences of generalized user characteristics and for testing a simple yes/no hypothesis relating to modal or
route choice, but is inadequate as a method of placing values on the
effect of user variables, or of qualifying the values for modal variables.
The hypotheses are not examined explicitly, only by inference.

The third approach assumes that one can directly relate the roles of
travel time and travel cost to each other in a choice context and that,
across a sample of individual travellers, a mathematical model can be
used to test an hypothesis on the implied trade-off between the underlying commodity characteristics. Earlier research (6, 11) identified two
alternative hypotheses that are central to this chapter:

Hypothesis 1:   Under habit conditions, typified by the journey to and
from work, the individual traveller only considers
potential substitution between alternative commodities
(modes) in his relevant choice set if his *usual* chosen
commodity (mode) becomes less attractive in terms of the
relevant characteristics of the alternative commodities
than it was previously. That is, the individual only substi-
tutes to make himself perceptually as well off as before.

Hypothesis 2:   Inferences about time values can properly be made only
by observing the behaviour of individuals with a 'trade-
off' between alternative options.

The remaining discussion in this section relates to the first hypothesis.
In the circumstances under which values of travel time savings are
estimated, it is assumed that the individual traveller has already selected
a mode, the 'usual' mode, and continues to repeatedly use this mode
until there is a large enough change in the level of the modal attributes
of this mode, or there is a change in the traveller's attitude towards the
characteristics of the mode, which prompts him to examine alternatives.
The situation in which the individual begins to consider the relative
advantages of alternative modal options can be appropriately referred to
as a point of potential substitution. That is the point at which the individual enters his decision space (11) and commences a search and learning
procedure which leads to a decision of whether or not to maintain his
usual habit mode or to select an alternative. To use this approach as
the basis for developing a method of directly valuing savings in travel
time (assuming, *ceteris paribus*, that we are strictly identifying only the
money cost of activity time) we need to be able to identify the level of
all the relevant commodity characteristics which define this position of
indifference (1). Since the levels of the relevant characteristics associated
with revealed preferences are, in themselves, not able to adequately
measure the point of potential substitution, it is also necessary to incorporate the underlying attitudes of individuals towards these modal

characteristics. This helps to gauge the dynamic role of the relevant characteristics and thus helps to define the level of the combination of relevant characteristics that places an individual in a situation in which he is choosing between alternative combinations of quantities of the given sets of characteristics which represent different modal options.

This approach involves the development of a modelling procedure which is concerned with 'individuals who use modes' (IM) in contrast to 'modes used by individuals' (MI). The former investigates current modal habits in terms of the various characteristics deemed relevant to the individual's selection of a travel mode. The individual (with a choice) is the unit of analysis. The latter orientation is essentially a statistical fitting process in which the modes (available to individuals) are the unit of analysis and the main aim is to produce a model of mode choice which maximizes the ability to predict future courses of action.

4.3.1. THE IM AND MI METHODS COMPARED

The major difference between the two approaches is the structure of the characteristics of the transport system. For the IM approach there is no reference to a particular mode, but to travel by a mode described as the 'usual' mode. In a situation of binary choice the competing mode is referred to as the 'alternative' mode.

If the object is to determine which modes are used by travellers, the traditional trade-off approach is appropriate (equation 1), with the explanatory variables expressed in terms of the value of the attribute for mode 1 (e.g. car) compared to the value of that attribute for mode 2 (e.g. train). When the emphasis is on the IM approach, the traditional choice modelling format is also inappropriate for statistical reasons. An example (Table 4.1) will help to explain this incompatibility.

For the usual/alternative dichotomy, the time difference and cost difference have the same sign, for a reversal of modes. The same value of y (the percentage of travellers using mode 1) is not consistent with the meaning of the dependent variable. This would place both modes at one end of the choice spectrum, negating any attempt to discriminate on the basis of time and cost. Hence any discrimination is hidden in a constant term. With the mode 1/mode 2 dichotomy, however, the sign changes conforming to the 1, 0 binary dependent variable. A set of parameters are required which will identify car users with a value as near to 1 as possible, and train travellers as close to zero. Observations are expected to cluster, as illustrated in Figure 4.1.

Around the point of indifference there is a cluster of observations with a non-definite committal to either mode. The IM approach can be readily illustrated as a pivot of Figure 4.1 around the point of indifference.

The new relationship is shown in Figure 4.2. Another dimension, not capable of being handled by the disaggregate binary choice model, is

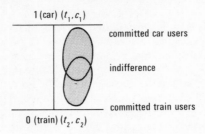

Fig. 4.1

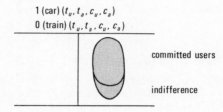

Fig. 4.2

now introduced.

The distinction between modes is lost, but a distinction about being committed and being indifferent is generated. It is quite possible to maintain a separate mode analysis by simply stratifying a data set by the ith usual mode, although the underlying choice hypothesis is now different from the assumptions of the traditional approach. This new distinction seems to be consistent with the development of a method of valuing travel time savings. The constant (α_0) now becomes a measure of inertia and is independent of mode. When all other things are zero, this is a measure of what the individual has to pay in order to change. We have now developed a series of arguments in building up an alternative approach to valuing travel time savings. In the next section these 'tools' will be combined to arrive at a testable valuation function.

4.4. THE VALUATION FUNCTION

Two definitions of the value of travel time savings are proposed to outline the difficulties of empirically determining the monetary equivalence of savings in time and the formulation of the valuation model:

Definition 1: The value of travel time savings is the amount of money an individual must be willing to outlay in order to receive a given amount of a composite characteristic named 'time' (of which time savings is only one element) and remain at the same level of satisfaction.

TABLE 4.1 *The relationship between the choice criterion, the IM method, and the MI method: a numerical illustration*

		Car	Train	IM method	MI method
Car user $(1, u, a)$	Time	20	30	$t_u - t_a = -10$	$t_1 - t_2 = -10$
	Cost	15	10	$c_u - c_a = +5$	$c_1 - c_2 = +5$
Train user $(2, u, a)$	Time	30	20	$t_u - t_a = -10$	$t_1 - t_2 = +10$
	Cost	10	15	$c_u - c_a = +5.$	$c_1 - c_2 = -5$

u = usual mode   1 = car mode
a = alternative mode   2 = train mode

Definition 2: The value of travel time savings is the amount of money an individual must be willing to outlay in order to receive a given amount of time savings and remain at the same level of satisfaction.

The major difference between the two definitions is the explicit consideration of the components of a composite characteristic 'time'. This highlights the variations in the circumstances under which travel takes place. Time savings is one component of the composite characteristic, the other components are referred to as the abstract summarizers 'comfort' and 'convenience' (9). Since comfort and convenience differentials are a function of changes in activity time, and are a function of a constant amount of activity time, the difficulties associated with empirically identifying a value of travel time savings according to the second definition become apparent. The non-time components need to be separated out, since when comparing transport projects a value of travel time savings is required that is not compounded at the comparison stage by the effects of the non-time components with magnitudes unique to each project. At the same time, it is necessary to develop appropriate comfort and convenience dollar equivalence rates so that the net benefits along those other physical dimensions for each project can be properly allowed for. While attempts to separate out variations in the circumstances under which travel time is spent have been made by the explicit introduction of additional variables, such as the probability of obtaining a seat, the qualitative nature of such variables and their high interdependency with travel time has limited their role in removing the variations in the circumstances under which travel time is spent. Attempts to identify and measure the extended physical dimensions of the abstract summarizers have not yet produced a definite solution to this issue (10). Hence any modelling procedure concerned with the derivation of a value of travel time savings at present accords essentially with definition 1. It is important that the definition is consistent with the values obtained from the model.

To utilize the concept of an indifference point with definition 1 we need to adjust the time-cost trade-off so that the individual is indifferent between modes. This requires a mechanism for overcoming the problem of discontinuity (associated with revealed preference) for each individual. In effect, this identifies the amount of cost change that would have to occur in a journey by the usual mode for the individual to consider an alternative mode of transport. This is an attitudinal concept. At this point an individual is potentially indifferent towards the two modes in terms of cost, time, and the circumstances under which time is spent. We choose to express this potential transfer payment in terms of a money outlay rather than a time outlay since it conforms with the definition of the value of travel time savings.

A model consistent with the above discussion is now proposed:

$$C = \alpha_0 + \alpha_1 X_1 + \alpha_2 X_2 + \ldots + \alpha_n X_n \tag{2}$$

where $C =$ the net monetary benefit of mode choice, equal
            to $C_u - C_a + TP_c$; TP_c is the transfer payment.

     $X_1 =$ the (perceived) time difference between the usual
            and alternative modes.

$X_2, \ldots X_n$ $=$ all other variables which are hypothesized (and measurable) as significant contributions to the perceived net benefit of the chosen mode.

The explicit model to be empirically tested is:

$$C_u + TP_c - C_a = \alpha_0 + \alpha_1 (t_u - t_a). \tag{3}$$

The coefficient of the explanatory variable gives a direct estimate of the value of travel time savings. The constant is a measure of inertia, independent of mode, representing the equivalent amount of money an individual must be willing to outlay in order to maintain indifference, for reasons other than those represented by the explanatory variables in the model.

4.5. SOME BASIC ADVANTAGES OF THE VALUATION MODEL

This model avoids the problems associated with the significance of the coefficient of monetized travel time savings derived from the ratio of two separate coefficients of travel time and travel cost. While separate coefficients might each be statistically significant, the ratio is often not significant. The coefficients of both variables may be significant, i.e. their standard errors may be small relative to the coefficients, but the ratio of these coefficients may have a large standard error. Lisco (15, p. 53) computed a standard error of $1·12 (U.S.A.) for the estimate of the value of travel time savings, producing a range almost as large as the estimate itself. There is no significance test for the ratio of two coefficients because we have no way of determining the degree of normality of the distribution (14, pp. 64–7).

Another issue is heteroskedasticity, commonly associated with regression constrained by a binary dependent variable. With a binary regressand, the stochastic error term is dependent on the expected value of the regressand, and hence the regressors, violating the classical least squares assumption of homoskedasticity, i.e. $E(e^2) = E(y)(1 - E(y))$. For a linear probability function, $y = X\beta + e$, since y is either 0 or 1, for a particular set of X's (X'), $e_t = -X'\beta$ or $1 - X'\beta$. For a zero expectation, its distribution must be

$$
\begin{array}{cc}
e & f(e) \\
-X'\beta & 1 - X'\beta \\
1 - X'\beta & -X'\beta
\end{array}
\quad \text{, and the variance }
\begin{aligned}
E(e^2) &= (-X'\beta)^2 \ (1 - X'\beta) \\
&\quad + (1 - X'\beta)^2 \ (X'\beta) \\
&= (X'\beta) \ (1 - X'\beta) \\
&= E \ (y) \ (1 - E(y)).
\end{aligned}
$$

Thus e varies systematically with $E(y)$ and hence with X.

In the modal choice model the function is modified to conform with the constraint of a binary regressand. In the new valuation model, on the other hand, the functional relationship is changed. By eliminating the binary constraint the heteroskedasticity associated with a binary dependent variable is removed.

4.6. THE INDIVIDUAL TRAVELLER VERSUS THE SITUATION

The second hypothesis of this chapter relates to the role of all travellers in a mode choice situation when the prime objective is the valuation of travel time savings. It is argued that individuals who face a dominant choice situation, with respect to the underlying travel time and cost characteristics influencing choice, and who choose an option not consistent with the rationality assumption (along the time and cost dimensions) must be making their choice on the basis of more than time and cost. These individuals face a *situation* with the property of illogical non-trading. The other groups of non-traders, the rational non-traders, behave consistently with what is expected, but they will behave consistently with any result obtained. In a sense their behaviour does not therefore contain any information about their decision criteria. This leaves the traders who face a situation where, in the two explicit characteristics space (with variations in the underlying circumstances in which time is spent), one mode is quicker and more expensive while the other is slower and less expensive. Traders versus non-traders is a function of the situation, not of the individual. In a sense we are all born traders. In valuing travel time savings the decision criteria require empirical analysis in terms of traders only. The identification of a trading situation is straightforward when only two characteristics are involved. It becomes more complex with the addition of further characteristics. In the present chapter we are only concerned with door-to-door travel times and travel costs. Inclusion of additional comfort and convenience dimensions, and the disaggregation of travel time into its heterogeneous components, has not yet been considered. This is an important area for future research.

The data from a recent study (6) can be used to illustrate the existence of traders and non-traders when only two characteristics are involved (Table 4.2). In the trader group, there are no time savers using the train. Out of the total sample, however, 170 individuals are time savers irrespective of the level of cost. Perhaps the most noticeable feature is the larger number of trading train users who are cost savers rather than time savers. Approximately 71 per cent of the sample are traders and hence qualified for incorporation in the model to empirically estimate the value of travel time savings.

TABLE 4.2   *The extent of traders and non-traders in a mode choice study (Sydney, Australia, 1971)*

	Situation	*Chosen mode*	*N*	*%*
Traders	time savers cost sacrificers	car	84	19·3
	time sacrificers cost savers	train	218	50·2
	time sacrificers cost savers	car	6	1·4
	time savers cost sacrificers	train	0	0·0
Non-traders	time sacrificers cost sacrificers	car	14	3·2
	time savers cost savers	train	25	5·7
	time savers cost savers	car	71	16·3
	time sacrificers cost sacrificers	train	17	3·9
Total			435	100

4.7. THE EMPIRICAL STUDY

The use of the valuation function to obtain a value of travel time savings for the journey to work required the collection of a new data set. The main objective in data collection was not to develop a model which would directly indicate the effect of policy changes on over-all modal split, but would enable a model of individual choice behaviour to be developed and tested in terms of factors whose inclusion in the model could be justified on the basis of a plausible hypothesis of such behaviour. The ideas discussed in this chapter were initially developed in 1971 when an initial set of data was obtained. A follow-up study was subsequently undertaken in 1973 to provide a basis for further testing the method. The basic results from the model for both years will first be discussed, followed by a more detailed investigation of the comparative results of alternative valuation procedures undertaken on the 1973 data set. The other important empirical matter is the reliability of the transfer payment results generated by the hypothetical question. The results from both data sets will be given, but the 1971 results will be discussed in more detail because of the unique opportunity for assessing the empirical results in the light of further empirical evidence related to the effect of a 50 per cent public transport fare increase, which occurred three months after the initial 1971 data had been collected (17).

The 1971 data set consists of a sample of commuters randomly selected, subject to a number of restrictions, from an area approximately 15 miles north of the Sydney Central Business District. The restrictions applied in sampling were:

1. The individuals must be commuters who use one of the main modes available for the greater part of the journey to work, and who have an effective choice between alternative modes of transport for that particular journey.
2. Each individual must have a fixed workplace address.
3. The household must own at least one car.
4. The individual must hold a current driver's licence.
5. People who require a car for use during working hours, or who have a company car for use in the journey to work, must be excluded on the grounds that their choice is not real.

The 1973 sampling frame was determined by the 1971 conditions, with allowance for replacement sampling because of mobility, non-response, and other intervening influences. The breakdown of the total sample of usable observations for 1971 and 1973 is given in Table 4.3. The usable response rate in 1971 was 70 per cent; in 1973 it was 58 per cent. The major reasons for the drop in the usable response were the annoyance associated with having to complete a form similar to the 1971 question-naire and the general saturation of the Sydney area in recent years by market researchers.

4.8. EMPIRICAL ANALYSIS

4.8.1. THE HYPOTHETICAL SITUATION

The transfer payment (TP_c) associated with a single one-way work

TABLE 4.3   *The samples of commuters—1971 and 1973*

	Main mode	1971		1973	
		N	%	N	%
Original Respondents	train	264	61·1	66	49·6
	car	173	38·9	67	50·4
	total	435	100	133	100
New Respondents	train	—	—	42	40·8
	car	—	—	61	59·2
	total	—	—	103	100
Total	train	264	61·1	106	45·3
	car	173	38·9	128	54·7
Grand total		435	100	234	100

trip, assuming ten work trips per week, was obtained from the following hypothetical question:

By what weekly amount would the TOTAL COST of your usual journey to and from work have to increase to make you consider using the best available alternative means of transport (which you can afford)?

The actual distribution of the transfer payment for the total sample and for the train travellers is given in Table 4.4 for both data sets. In order to check the relative validity of the magnitudes of the transfer payment, a comparison can be undertaken of the 1971 results and the results of a study reporting on a rail fare increase that occurred three months after the initial data was collected (17). Fare charges before and

TABLE 4.4   *Distribution of the transfer payment—1971 and 1973*

Weekly transfer payment	Total sample			Train travellers			Total sample cumulative %	
	1971		1973	1971		1973	1971	1973
	N	%	%	N	%	%		
less than 50 cents	41	9·4	7·2	28	10·6	11·0	9·4 >	7·2
50 cents—99 cents	33	7·6	6·8	23	8·7	7·3	17·0 >	14·0
$1·00—$1·49	42	9·7	10·6	23	8·7	10·1	26·7 >	24·6
$1·50—$1·99	41	9·4	7·2	30	11·4	12·8	36·1 >	31·8
$2·00—$2·49	71	16·3	11·0	49	18·6	10·1	52·4 >	42·8
$2·50—$2·99	49	11·3	5·9	33	12·5	5·5	63·7 >	48·7
$3·00—$3·49	37	8·5	8·9	20	7·6	6·4	72·2 >	57·6
$3·50—$3·99	10	2·3	3·4	6	2·3	3·7	74·5 >	71·0
$4·00—$4·49	13	3·0	4·2	7	2·7	2·8	77·5 >	75·2
$4·50—$4·99	2	0·4	2·5	2	0·8	2·8	77·9 >	77·7
$5·00 and over	96	22·1	32·3	43	16·3	27·5		
Total	435	100	100	264	100	100	100·0	100·0

TABLE 4.5 A sample of rail fares before and after the July 1971 increase

Station	Single fare			Return fare			Weekly fare			Yearly fare		
	Before [cents]	After	Change %	Before [cents]	After	Change %	Before [cents]	After	Change %	Before [cents]	After	Change %
Asquith	47	70	33	92	150	39	269	420	36	13 700	19 730	31
Hornsby	42	62	33	82	122	33	266	400	34	12 215	18 790	35
Pennant-Hills	50	75	33	100	150	33	260	425	39	12 220	19 970	39
Waitara	42	60	30	82	122	33	250	400	38	13 200	18 850	30

after the fare increase, for a sample of journeys from four of the railway stations in the Study Area to the Sydney Central Business District, are shown in Table 4.5. The sample of train travellers in March 1971 were re-interviewed on two successive occasions, in mid-October 1971 and in February 1972. The results of the first re-interview can be used to analyse the reliability of the transfer payment approach.

176 of the 264 initial 1971 travellers adequately completed and returned by mail the October post fare increase questionnaire. Out of these 176 ex ante train travellers who used rail for the greater part of their journey to work, 87·5 per cent remained train travellers after the July fare increase. The relationship between the distribution of the transfer payment and the fare increase is summarized in Table 4.6. A more useful indication of the consistency between the transfer payment result, the fare increase, and the potential or actual modal switching is given in Tables 4.7 and 4.8. The information was obtained on an individual basis and then aggregated. The interpretation of Table 4.7 is, for example, that 2·6 per cent of the individuals who continued to travel by train after the fare increase indicated prior to the fare increase that the cost would have to increase, on average, by 7·5 cents per trip for them to *consider* using the alternative means of transport. Of the ex post train travellers, however, 12·3 per cent incurred an average fare increase of 7·5 cents per trip. The mean difference between the transfer payment and the fare increase for the ex ante and ex post train travellers is plus 9·8 cents per trip which, given the tendency to round the total cost of travel to the nearest 5 cents (6), gives confidence to the use of the transfer payment as a guide on attitudes to cost changes. The mean difference for ex ante train travellers who are ex post car travellers is minus 1·4

TABLE 4.6   *The distribution of the transfer payment and the fare increase. Comparison sample, March and October 1971*

Weekly transfer payment and fare increase	N (TP_c)	N (fare increase)	% (TP_c)	% (fare increase)
less than 50 cents	19 >	3	10·8	1·7
50 cents–99 cents	12 <	18	6·8	10·2
$1·00–$1·49	17 <	62	9·7	35·2
$1·50–$1·99	17 >	57	9·7	32·3
$2·00–$2·49	37 >	12	21·0	6·8
$2·50–$2·99	24 >	8	13·6	4·5
$3·00–$3·49	14 >	8	7·9	4·5
$3·50–$3·99	4 >	2	2·3	1·1
$4·00–$4·49	5 <	1	2·8	0·5
$4·50–$4·99	1 <	4	0·1	2·2
$5·00 and over	26 >	1	14·8	0·5
Total	176	176	100	100

TABLE 4.7   *The relationship between the transfer payment, the fare increase, and the extent of modal switching. Comparison sample, March and October 1971 (N = 176)*

Weekly transfer payment and fare increase	Ex post train traveller				Ex post car traveller			
	$TP_c > fare\uparrow$		$TP_c < fare\uparrow$		$TP_c > fare\uparrow$		$TP_c < fare\uparrow$	
	TP_c	$fare\uparrow$	TP_c	$fare\uparrow$	TP_c	$fare\uparrow$	TP_c	$fare\uparrow$
Less than 50 cents	0·0	1·7	35·0	0·0	0·0	11·1	30·8	0·0
50 cents – 99 cents	2·6	12·3	17·5	0·0	11·1	44·4	7·7	0·0
$1·00 – $1·49	4·4	38·6	22·5	30·0	11·1	22·2	7·7	30·8
$1·50 – $1·99	12·3	33·3	5·0	43·0	0·0	11·2	15·4	7·7
$2·00 – $2·49	22·7	5·3	12·5	7·0	33·3	11·1	23·0	15·4
$2·50 – $2·99	16·7	3·5	2·5	5·0	33·3	0·0	7·7	15·2
$3·00 – $3·49	8·8	4·4	5·0	5·0	11·2	0·0	7·7	7·7
$3·50 – $3·99	3·5	0·0	0·0	5·0	0·0	0·0	0·0	7·7
$4·00 – $4·49	4·4	0·9	0·0	0·0	0·0	0·0	0·0	0·0
$4·50 – $4·99	0·9	0·0	0·0	5·0	0·0	0·0	0·0	15·5
$5·00 and over	23·7	0·0	0·0	0·0	0·0	0·0	0·0	7·7
Total	100	100	100	100	100	100	100	100

TABLE 4.8   *The relationship between the transfer payment, the fare increase, and the mode of transport. Comparison sample, March and October 1971 ($N = 176$)*

		Train travellers		Alternative mode travellers		Total	
		N	*%*	*N*	*%*	*N*	*%*
$TP_c \leq$ fare increase	*N*	40		13		53	
(should *consider* using	*%*		75		25		100
alternative mode)		26		59			
$TP_c >$ fare increase	*N*	114		9		123	
(should *not* consider	*%*		93		7		100
using alternative mode)		74		41			
Total		154	100	22	100	176	100

cents, which is also directionally consistent.

Out of all the individuals who faced a fare increase that exceeded the amount by which cost would have to increase for them to consider an alternative mode for the journey to work, only 25 per cent actually changed mode. Seventy-five of the individuals would have apparently considered an alternative mode but did not change, perhaps for non-cost reasons. The wording of the hypothetical question did not make any explicit allowances for potential non-cost gains or losses associated with a switch.

For the individuals facing a fare increase which was less than the amount by which the cost would have to increase for them to consider an alternative mode, 93 per cent conformed to what was expected; namely, maintenance of their present modal pattern, although 7 per cent actually switched to the car for the journey to work. The ex ante over-statement of some individuals' elasticity of substitution between modes possibly stems from the nature of the question on the transfer payment and any other non-cost changes which might have occurred during the period between March and October 1971. In accordance with the expected relationship between the transfer payment, the fare increase, and the mode used subsequent to the fare increase, 72 per cent of the travellers were definitely responding in the same direction as the ex ante hypothetical question had suggested. The remaining individuals responded by either switching mode, for reasons apparently not associated with the indicated cost relationship (although we have no evidence to distinguish between perceptual error in supplying information on the transfer payment or any other non-cost explanation), or they considered the alternative mode but did not change.

For the 9 individuals who switched to an alternative mode the average cost difference between the transfer payment and the fare increase was 11 cents. This is small enough to be accounted for by variations due to the passage of time (between the survey and the fare change) and other perceptual adjustments. In cannot be argued that these respondents (28 per cent of the total) place the hypothetical question in doubt, since influences during the intervening period must have occurred. The 72 per cent who were correctly classified, on the other hand, do give some initial confidence in the use of the hypothetical question. The general directional tendencies appear to support the hypothesis.

4.8.2. THE VALUATION MODEL: BASIC RESULTS

The results of the basic model, estimated on both data sets, are summarized in Table 4.9. The models are statistically sound with highly significant t-values associated with the estimates of the parameters of the explanatory variable. The values of travel time savings associated with the trader sample and derived directly from the model are 57 cents per person hour for the 1971 data set and 50 cents per person hour for the 1973 data set. The comparable 1971 total sample result is 24 cents per person hour, nearly 60 per cent lower than that for the traders. Since the value of travel time savings is usually expressed as a percentage of the average wage rate, this was used as a basis for updating the value of travel time savings (without questioning the appropriateness of the usual assumption about the role of income in valuing travel time). The equivalent values are 27 per cent of the average wage rate for the 1971 data set and 14 per cent for the 1973 data set. A major advantage of the above valuation model is that the standard error of the value of travel time savings is given directly by a single standard error and by an accompanying confidence level. At a 95 per cent level of confidence, the standard error associated with the mean value of travel time savings is ±10·05 cents per person hour for 1971 and ±14·84 cents per person hour for 1973. The corresponding 1971 total sample result is ±7·5 cents per person hour.

The relatively stable absolute value of travel time saving between the two years contrasts with the significant decline in the value as a percentage of the average wage rate. Between March 1971 and June 1973 the average annual income per employed person increased by $2750, and the general rate of inflation increased from 4 per cent to 17 per cent per annum. This significant increase in annual income of 65 per cent raises the question of the appropriateness of the basis for expressing the value of travel time saving. When incomes have increased at a high annual rate, which is not greater than changes in the consumer price index, the equivalence interpretation of the value of travel time savings is likely to lead to an over-estimate of the value when the basic equivalence percentage (usually 25 per cent for in-vehicle time, and a lower percentage

TABLE 4.9   *The basic valuation results—1971 and 1973*

$$C_u + TP_c - C_a = \alpha_0 + \alpha_1(t_u - t_a)$$

	Parameter	Estimate of parameter	Standard error of estimate	t-value	Level of significance	R^2	Mean annual income ($)	Year of data collection	N	$t_u - t_a$	$C_u - C_a$
Traders	α_1	−0·9566	0·0855	−11·18	1%	·295	4269	1971	301	8·8	−16·9
	α_0	−18·9983									
	α_1	−0·8289	0·1262	−6·57	1%	·268	7020	1973	120	−3·9	−11·5
	α_0	−18·6450									
Traders and non-traders	α_1	0·3963	0·0642	−6·18	1%	·274	4285	1971	435	3·4	−15·8
	α_0	−13·2157									

for door-to-door time) is adjusted annually to allow for increases in the real value of travel time and inflation.

The above results generally support the argument that the absolute value of travel time savings is relatively stable over a two-year period. It seems more desirable to also relate adjustments in the value of travel time savings to changes in the consumer price index. Perceived spending power has decreased in recent years together with changes in the general attitude towards spending money on transport facilities. The transfer payment is a useful concept in helping to understand the relationship between attitudes to outlays of money and the potential gains associated with monetary outlays. The transfer payment attempts to reflect the role of the individuals' value structures in trading off money costs against other things. From Table 4.4, at any cumulative level, the percentage of 1971 respondents exceeded the percentage of 1973 respondents. This suggests that the increase in travel cost in 1973 which persuaded a person to consider an alternative mode was greater than that required in 1971. The mean transfer payment in 1971 was 26·5 cents per one-way trip in contrast to 36·8 cents in 1973. This represents an increase of 10·3 cents, or 39 per cent. Since mean income increased by 65 per cent it could be inferred that the income elasticity of demand is less than one and that the value attached to money in the context of travel choice is increasing at a rate greater than the increase in income. Hence we could expect the value of travel time savings to fall slightly (in absolute units), but not by as much as the equivalence measure suggests.

Since 1971 it is quite possible that individual's preferences might have changed, even though the role of travel time itself might have remained stable. It might be reasonable in mode choice studies to interpret the constant term as something that is fixed to give the correct calibration; for studies concerned with the value of travel time savings, on the other hand, the constant term has to be more carefully interpreted. Indeed, differences in estimates of the value of travel time savings might be attributable to this cause. The value associated with the constant term for the traders has remained unchanged between the two years. The small difference in the value of travel time savings appears to be related to changes in the attitude towards the trade-off between travel costs, travel time, and the circumstances under which travel time is spent.

From the above discussion, it seems possible that a value of travel time savings equal to 14 per cent of the average wage rate might be reasonable. There will naturally be doubt about this result, because of the general acceptance of a door-to-door value as a percentage of the average wage rate within the range 20 per cent to 25 per cent. In the absence of any previous research into empirical values of travel time savings, it is possible that the 14 per cent result would not even be questioned. We are suggesting that the value of travel time savings should in future be related to changes in the consumer price index and in the average wage rate. The general level of prices *vis-à-vis* income levels does have an effect on attitudes towards the allocation of money to particular activities. An annual constant percentage mark up on the value of travel time savings, expressed as a percentage of the average wage rate, to 'allow for' inflation seems to over-estimate the willingness to outlay money to save travel time and any other associated gains.

In concluding this section it is useful to compare the 1973 value of travel time savings associated with the traders with the value derived from the traditional approach. The value of travel time savings for the trader set is 85 cents per person hour, equivalent to 24·3 per cent of the average wage rate. Once again, the derivation using the traditional approach would support the more 'acceptable' value, (nearly 25 per cent of the average wage rate). The methodology of the alternative valuation model seems to provide more scope for assessing the appropriateness of the values obtained; provided the first hypothesis in section 4.3. is accepted as at least as plausible as the hypotheses underlying the traditional approach.

4.8.3 APPLICATION OF THE RESULTS

To provide some guidance on the selection of a value of time savings for predicting demand, the values of travel time savings obtained from the 1973 data set, using both the traditional and the valuation models, were used to construct a generalized cost variable, which was then used to assess the predictive capability of the mode choice model. The expected number approach was used on a 50 per cent random separation of the data set (9, p. 41). The expected value of a random variable is the sum of the products of the potential outcomes and the probabilities of their occurrence. The mode choice model (equation 1) is calibrated using a randomly selected sample of approximately 50 per cent of the total sample. The remaining sample observations are then applied to the calibrated model to assess the predictive capability. This procedure was applied to both sample sub-sets. The results are summarized in Table 4.10. A comparison of the empirical results does not provide any basis for selecting between the two values of travel time savings. The predictive capability of the model is insensitive to changes in v within the range tested.

While the arguments developed to obtain values of travel time savings are an important ingredient in the basis for recommending a range of values, it appears from the above evidence that the methodological arguments are much more important for the selection of values relevant to the calculation of users benefits for purposes of economic evaluations rather than for developing a generalized cost variable for mode choice prediction.

TABLE 4.10 *The predictive capability of the mode choice model with alternative values of travel time savings (v), 1973. Total sample N = 236*

Choice of mode $= \alpha_0 + \alpha_1 ((C_1 - C_2) + v(t_1 - t_2))$

Origin of V	v cents/hour	Sub–sample A			
		N (for calibration)	N (application)	N (actual car users)	N (predicted car users)
Traditional approach (Traders)	85	119	117	61	61·2
Valuation function (Traders)	50	119	117	61	61·1
		Sub-sample B			
Traditional approach (Traders)		117	119	67	67·4
Valuation function (Traders)		117	119	67	66·8

4.9. CONCLUSION

The alternative approach to valuing travel time savings has attempted to raise the whole issue of the possibility of alternative hypotheses that might relate to the underlying decision processes leading to choices under existing conditions of habit. The application to commuter travel behaviour seems consistent with the assumptions of the procedure. The next research requirement is to extend the model to allow for variations between the various heterogeneous components of travel time and to consider its relevance in the context of other journey purposes. The empirical analysis leads us to conclude that the value of door-to-door commuter travel time savings is 50 cents per person hour. However, in the evaluation of road investments, values of time savings should be used that are consistent with the type of time savings involved (i.e. in-vehicle, waiting, or walking). The reader is left to judge whether this alternative procedure is more consistent than the traditional approach with the decision processes underlying the choice of mode for the journey to work.

REFERENCES

1 Boulding, K. E., The Formation of Values as a Process in Human Learning, *Transportation and Community Values*, Highway Research Board Special Report, no. 105, Washington, D.C., March 1969.

2 Commonwealth Bureau of Roads, *Economic Evaluation of Road Improvements from the Australian Roads Survey, 1967–68: Procedures and Results*, Associated Bureau Papers, no. 21, Commonwealth Bureau of Roads, Melbourne, 1969.

3 Davies, A. L. and Rogers, K. G., *Modal Choice and the Value of Time*, LGORU Rpt. C143, 1973

4 Delofski, E. F., *Route Choice and the Value of Inter-Urban Travel Time Savings*, unpublished B.Com. (Hons.) Thesis, School of Economics, University of New South Wales, Australia, December 1973.

5 Haikalis, G. and Hyman, J., Economic Evaluation of Traffic Networks, *Studies in Highway Engineering Economy*, Highway Research Board Bulletin 306, Washington, D.C. 1961.

6 Hensher, D. A., The Consumer's Choice Function: A Study of Traveller Behaviour and Values, unpublished Ph.D. thesis, School of Economics, University of New South Wales, Australia, October 1972.

7 Hensher, D. A., Valuation of Travel Time: An Alternative Procedure, paper presented at the *Third Australasian Conference of Economists*, University of Adelaide, South Australia, May 1973.

8 Hensher, D. A., Review of Studies Leading to Existing Values of Travel Time, *Valuation of Travel Time*, Transportation Research Board Special Report, Washington, D.C., 1975.

9 Hensher, D. A., Mcleod, P. B., and Stanley, J. K., Comfort and Convenience: Theoretical, Conceptual and Empirical Extensions of Disaggregate Mode Choice Models, Melbourne, Australia, March 1974, (unpublished).

10 Hensher, D. A. and Mcleod, P. B., Generalizing and Generalized Cost Function, Melbourne, Australia, December 1974 (unpublished).

11 Hensher, D. A., Perception and Commuter Mode Choice: An Hypothesis, *Urban Studies*, 50, February 1975.

12 Hoch, I., Benefit Cost Method for Evaluating Expressway Construction, *Traffic Quarterly*, April 1961.

13 Lee, N. and Dalvi, M. Q., Variations in the Value of Travel Time, *Manchester School*, 37, no. 3, pp. 213–36, 1969.

14 Lianos, T. P. and Rausser, G. C., Approximate Distribution of Parameters in a Distributed Lag Model, *Journal of American Statistical Association*, 67, no. 337, March 1972.

15 Lisco, T. E., The Value of Commuters' Travel Time: A Study in Urban Transportation, unpublished Ph.D. Thesis, Department of Economics, University of Chicago, 1967.

16 Marshall, A., *Principles of Economics*, 8th edn., London.

17 Paterson, J. and Hensher, D. A., *Elasticity of Demand for Urban Public Transport*, John Paterson Urban Systems, Melbourne, July 1972.

18 Weiner, E., The TRANS Urban Model System and its Application to the National Transportation Study, Paper presented at the International Conference on Transportation Research, Bruges, Belgium, June 1973.

5

The Value of Travel Time Savings and Transport Investment Appraisal *

by A. JENNINGS AND C. SHARP

5.1. INTRODUCTION

The purpose of this chapter is to examine the influence of the values given to travel time savings on investment decisions. As is shown in other chapters of this book many of the problems of evaluating time savings are still unsolved. We have attempted to identify those issues which appear to be of the greatest immediate importance in affecting decisions about the allocation of resources to transport projects. In particular, we have looked at the problems which may result from the practice of using a few average figures to represent the time saving values of different classes of travellers, transport mode, and types of traffic. Most, but not all, of the discussion relates to investment in roads.

It is apparent both that the amount of utility derived from time savings and their money value will be influenced by a number of variables and that a wide range of values may be obtained. A young man hurrying to meet his girlfriend may feel that he benefits greatly from a five-minute reduction in his journey time, while the same time saving might have little value to a Sunday-afternoon pleasure motorist. The unit value of time savings measured by what consumers are willing to pay for them is likely to be influenced by the level of income they enjoy. Some forms of travel are more disliked than others, so that, for any individual, journey time reductions on the preferred mode may be of less value than those on the inferior mode. Time saving values may not be the same for both goods and passenger traffic and may vary with the size of the saving. The influence of all these variables suggests that individual or group time saving values may be widely dispersed in relation

* This chapter is a revised version of a paper which first appeared in the University of Leicester Economics Discussion Papers series. The names of the authors have been placed in alphabetical order.

to their average values.

Since time savings usually form the major part of the benefit of transport investment a knowledge of their money value is clearly of considerable interest. The problem of finding meaningful money values for time savings is particularly important in evaluating investment in transport infrastructure which is normally provided without any direct charge to consumers. Whereas a mistake in estimates of how much people will be prepared to pay to enjoy the speed advantages of supersonic flight may eventually be revealed wrong estimates of the value of the benefits from a new road may never be detected. Repeated misallocation of investment between transport and other sectors of the economy, or between alternative transport investments, may result from the use of faulty estimates of the value of time savings.

The theory of time valuation is reviewed very briefly in section 5.2. and empirical evidence on time savings values is discussed in 5.3. with a summary of some measured and recommended values in Table 5.1. Reference to new work described in other chapters of this book is excluded. In section 5.4. problems relating to the influence of journey purpose, consumers' income, journey mode, type of traffic carried, and the size of time savings are discussed. In section 5.5. some of these problems are related to the appraisal of an actual road investment project, the Granby Halls scheme in Leicester. Finally, in section 5.6. the main policy implications of the discussion are set out.

5.2. CURRENT THEORY OF TIME VALUATION

Recent discussion on the value of time (1) has shown that the current practices of valuing working time on the basis of wage rates and of devising general time values from behavioural models of travel choice are unsatisfactory, and this dissatisfaction is reflected in the uncertain state of the theory of the value of time. Some writers reject the traditional theory of the value of time since its prediction that travel time is valued at the rate of pay is not confirmed by the evidence. Attempts have been made to develop a general theory of the value of time based on neoclassical consumer theory in which conditions are found for maximizing utility when a consumer has to chooce between alternative forms of consumption and is subject to a budget constraint. G. Becker (2) argues that the full costs of goods should include both market prices and the cost of time used in their consumption. Time can be converted into goods by working longer hours and foregoing time spent in consumption. The total resource constraint in Becker's utility maximizing model becomes the maximum money income that could be earned if the minimum possible time were devoted to any activity except work. Alan Evans (3) highlights the confusion which has arisen in some theoretical work between the value of change in the total time available and the value of being able to spend time in any particular activity. As regards the total supply of time, it is doubtful if this can be changed in any meaningful way, since no one can add a twenty-fifth hour to the day, although in certain circumstances a man may be able to prolong his life. Evans uses a utility maximizing model in which an individual can allocate his time between costless activities (e.g. going for a walk), activities for which he is paid, and those for which he must pay and which are subject only to a total time constraint and to a budget constraint. The utility maximizing position can be represented by the equation:

$$u_i - \lambda r_i = \mu \ (i = 1 \dots n)$$

where: u is the individual's marginal utility from spending time in the ith activity;

λ represents the marginal utility obtainable from relaxing the budget constraint;

μ is the marginal utility from relaxing the time constraint;

r_i represents the payment made or received for the ith activity.

Where a consumer has allocated his time optimally a marginal saving of travel time would have zero value. If the consumer transfers a small amount of time from travelling to working then he is exactly compensated by the payment made to him (i.e. r_i is negative in the model); if he uses the time saving for a leisure activity for which he must pay then the utility from this activity is exactly balanced by the payment that he makes. This result, whereby the consumer is neither better nor worse off, as Evans points out, is clearly in conflict with the empirical evidence that people behave as if a travel time saving made them better off. This contradiction may be explained by the fact that the assumption in the model that people are able to allocate time in very small amounts to any activity is erroneous. Both De Serpa (4) and Becker (2) have also shown how certain activities may require a minimum consumption of time. Thus Evans argues that 'time savings are only of value if the individual can transfer the time saved to some other activity which he prefers' (13).

5.3. EMPIRICAL EVIDENCE

As yet there is no agreement on the time values which have emerged from empirical studies. The lack of consistency in results, shown in Table 5.1, arises because of the number of variables to be considered; difficulties in selecting accurate samples; variance in actual preferences; discrepancies between perceived and calculated values of explanatory variables; and also because basic assumptions may be unsound.

5.3.1. VALUE OF TIME AND JOURNEY PURPOSE

As work times have been derived from wage rates this explains why

TABLE 5.1   *Summary of time savings values at 1968 U.K. prices*

Source	Year of data collection or investment appraisal	Mode		Value of unit of time saved (per hour/person)
(a) Commuter and non-work time:				
Merlin and Barbier (17)	1961	PT, Car Car, PT		27 48
Victoria Line Study (20)	1962	–		31
Quarmby (6)	1966	Car/Bus	minimum value maximum value	7 25
Beesley (5)	1963	Public transport Car	minimum value maximum value minimum value maximum value	12 19 12 25
LGORU (7)	1967/8	Average of bus/train/car	combined data Liverpool Manchester Leicester Leeds	43 39 32 84 22
Lee and Dalvi (16)	1966	Car PT		66 40
Edinburgh Study (11)	1969 1969	Train Car		41 59
Solent Travel Study (9)	1970	Hydrofoil, Hovercraft versus Ferry Less than} 25 km}	All journeys/ minimum maximum minimum maximum	34 39 11 12
Dawson and Everall (15)	1969	Car	small medium large	26 53 119
Henscher and Hotchkiss (17)	1970	Ferry		17
Roskill (13)	1968		minimum maximum	11 34
Commonwealth Bureau of Roads (21)	1974	Car Bus		11 11
Department of Environment (22)	1974	In-vehicle walking & waiting		17 34

TABLE 5.1 *continued*

Source	Year of data collection of investment appraisal	Mode	Value of unit of time saved (per hour/person)	
(b) Pure leisure time:				
Dawson and Smith (12)	1957	Car	minimum value maximum value	50 68
Solent Travel Study (9)	1970	Hydrofoil Hovercraft versus Ferry		negative
Dawson and Everall (15)	1969	Car	medium large	72 75
Channel Tunnel Study (14)	1973	–		16
(c) Work time:				
Victoria Line Study (20)	1962	–		43
Solent Travel Study (9)	1970	Hydrofoil, Hovercraft versus Ferry	All journeys less than 25 km minimum } maximum	negative 63 72
Dawson and Everall (15)	1969	Car	medium large	201 298
Roskill (13)	1968	–	minimum maximum	*146* *258*
Channel Tunnel Study (14)	1973	–		*70*
Commonwealth Bureau of Roads (21)	1974	Car		*146*
		Light truck		*71*
		Articulated vehicle		*90*
Department of the Environment (22)	1974	All workers		*128*
		Car drivers		*146*
		Car occupants		*127*
		Rail users		*157*
		Bus users		*77*
		London Underground users		*139*
		Light goods vehicle driver		*64*

TABLE 5.1 *continued*

Source	Year of data collection of investment appraisal	Mode	Value of unit of time saved (per hour/person)
(c) *Work time*:			
		Light goods vehicle occupant	55
		Heavy goods vehicle driver	70
		Heavy goods vehicle occupants	58
		Bus drivers	72
		Bus conductors	66

Note: The italicized figures are theoretical values, in that they have been estimated from earnings. Other figures have been derived from observation of the preferences of travellers in a situation of choice. All values were converted into sterling and recalculated on a constant price basis with 1968 = 100. The Index of retail prices for 'All Items' was used to adjust values, subsequent indices being linked without making any allowance for changes in weights.

PT = Public Transport

the greater number of studies of time saving have related to commuter time. Beesley (5) tried to find that value of time which best explained the observed choices of a sample of employees working in the Ministry of Transport in London, by minimizing the number of people who apparently were making bad choices—i.e. who by choosing the other mode could have either saved time by paying less money than this value of time implied, or saved money at the expense of a smaller time increase than the value of time implied. Quarmby (6) used a more elaborate discriminant analysis model to explain the choices made between car and public transport by a sample of car owners working in central Leeds. The model took explicit account of a number of other factors, besides over-all travel time and costs, giving a purer trade-off between cost and travelling time. The data collected in Leeds has been incorporated into more extensive studies of factors determining modal choice and of time values by LGORU, using discriminant analysis and 'limiting time value' techniques (7). Most of the LGORU figures relate only to commuters, but a study of travel between Edinburgh and Glasgow included data on trips taken for purposes other than the journey to work. An analysis of the Edinburgh/Glasgow data suggested that people valued time on commuter trips and on work journeys at a lower rate than time on shopping or recreational trips (but see argument in 5.4.1. below).

A technique called 'priority evaluator' (8) has recently been used to estimate the importance which commuters attach to time savings. Some London commuters were asked about their existing commuter journey and, as part of the priority evaluator method, were also asked to allocate a given travel 'expenditure' in an optimal fashion among a range of priced alternatives. These included walking, waiting, and travelling time; convenience; seating; crowding or congestion; reliability; and journey cost. The important variables selected were comfort and convenience, with travelling time receiving a lower value. It must be emphasized that the ranking is a function of the relative 'prices' assumed.

Evidence on the value of the time savings of those who are neither working nor commuting is at present very limited. Leisure trips may provide some utility in themselves as well as being inputs to other activities. Both the Solent Crossing Study (9) and a study of short urban leisure trips to libraries in Birmingham (10) gave negative time values, although it may be that the comfort and convenience of the slower route explains this phenomenon. Watson's Edinburgh/Glasgow data (also used by LGORU) gave positive values of 43·0p per hour for train journeys and 62·5p for car journeys made in leisure time (11).

An earlier study of leisure travel, based on trade-offs between time and cost in the choice of route, gave a minimum likely value for time of 36p per hour per motorist and a maximum of 49p per hour per motorist (12). In the evidence submitted to the Roskill Commission (13) Professor N. Lichfield argued that leisure time should be valued at only 2·5p per hour, while others appeared to believe that leisure time savings had no value at all for air passengers en route for the airport. In the Final Report, non-working time was valued at 22·5p per hour, or 25 per cent of the average income of those 'leisure' travellers interviewed in a sample survey taken at Heathrow and Gatwick airports. In the generalized cost function of U.K. holiday passengers used in the Channel Tunnel Study a time cost of 23p per hour was used, based on figures prepared by the Department of the Environment and adjusted to allow for increases in real income (and hence in the value of time) between 1972 and 1980 (14). A recent study has been made by the T.R.R.L. (15) of the value of motorists' time in Italy, using logit analysis to model route choice. Data were collected from autostrada and competing free all-purpose roads. Even given correction for vehicle occupancy, the values were high relative to the values used currently by the Department of the Environment (D.O.E.). There are several possible reasons why the autostrada values are higher than those found in Britain. Although a route choice study eliminated the distortion of the 'comfort factor' between routes. It could be that the qualitative difference of travelling on motorways, compared to all-purpose roads, are not as great in Britain. It could also be that speed has a higher subjective value to Italians that it does to Britons.

TABLE 5.2 *Value of time and income*

Source	Year of data collection or investment appraisal	Type of time	Value of time as % of hourly income		
Beesley (5)	1963	Commuter			30−50
Quarmby (6)	1966	Commuter			20−25
LGORU (7)	1967/8	Commuter	61% combined data		
LGORU (7)		Commuter	61% Liverpool		
LGORU (7)		Commuter	47% Manchester		
LGORU (7)		Commuter	132% Leicester		
LGORU (7)		Commuter	30% Leeds		
Lee & Dalvi (16)	1966	Commuter			15−45
Roskill (13)	1968	Work			149
		Leisure			25
Dawson & Everall (15)	1969	Commuter & non-work			75
			Small car	Medium car	Large car
		In course of work	−	303	225
		To & from work	60	80	89
		Other	−	109	65
		All purposes	166	133	110

5.3.2 VALUE OF TIME AND INCOME

Most empirical studies assume a relationship between the valuation put on time savings and income levels. As the survey of empirical evidence in Table 5.2 indicates, the ratio varies widely. Thus Beesley found time savings to be valued at 30−50 per cent of hourly income, Quarmby at 20−25 per cent, and Lee and Dalvi (16) at 15−45 per cent. The general conclusion of the LGORU studies (in which the same models were used to explain modal choice and evaluate time) was that 'The value of an individual's travel time may be dependent on his income. No reliable numerical relationship was, however, obtained, (7.)

The value for commuting and other non-work journeys found in the Dawson and Everall autostrada study was approximately 75 per cent of the average wage rate. It is interesting to note that a French study of commuter time savings in Paris also gave 75 per cent of the wage rate (17).

5.3.3. VALUE OF TIME AND JOURNEY MODE

A criticism of earlier empirical work, for example Beesley's pioneering study, was that no differentiation was made between the alternative ways in which travelling time can be spent. Later studies found that savings in different types of travelling time are valued differently. People apparently dislike waiting for a bus or train or walking, more than they dislike travelling time spent in a car or bus. Quarmby found that some of the evidence suggested that car time might only be worth 40–50 per cent of bus time, but the statistical evidence is inconclusive. Values for walking, waiting, and in-vehicle times found in the LGORU study are shown in Table 5.3.

TABLE 5.3   *Value of time and journey mode*

| | (Pence/hour) | | |
	Walking time	Waiting time	In-vehicle time
Liverpool	65	35	21·5
Manchester	74	100·5	27·5
Leicester	150	160	60
Leeds	35	57·5	9·5

Source: LGORU (7)

In the Solent Crossing Study the modal choice explanation was improved by the addition of walking and waiting time variables. A study of short urban leisure trips to libraries in Birmingham gave negative time values until a 'comfort' proxy, car- or bus-ride time, was added to the mode choice equation. The priority evaluator study carried out by S.C.P.R. indicated that commuters in London attach weights to things like reliability and number of journey interchanges. The weights attached to the variables differed between modes, so that improvement priorities were mode-specific. Thus the value attached to an improvement in travel time would vary according to the mode involved.

5.3.4. VALUE OF TIME AND SMALL AND LARGE TIME SAVINGS

The only modal choice study so far to attempt to disaggregate the data by size of time saving is an analysis of the value of time for commuting motorists carried out by the Stanford Research Institute (18). A route choice model, based on the logit function, was used to predict the motorists' route choices between alternative toll and free roads. Their results showed a higher value of time for motorists with a higher income, and that for very small amounts of time saved the value of time per minute was quite low, increasing with the amount of time saved and then decreasing again. Thus the equivalent hourly values for a motorist with a family income of $9000 per annum were: $0·79 per person per hour for the first minute saved; $3·72 for the fourteenth minute saved; and $1·26 per person per hour for the twenty-fifth minute saved.

The Third London Airport Study assumed constant unit time values in all its main calculations, but in response to criticism examined the

TABLE 5.4   *Small and large time savings*

Hourly equivalents of the total value of time saved in dollars

Time saved in minutes	Income level of motorists							
	1	2	3	4	5	6	7	8
1	0·43	0·53	0·65	0·79	0·95	1·14	1·36	1·62
5	0·43	0·53	0·65	0·79	0·95	1·14	1·36	1·62
10	0·55	0·78	1·07	1·42	1·82	2·27	2·74	3·22
15	0·72	1·08	1·49	1·97	2·45	2·92	3·40	3·87
20	0·66	1·00	1·36	1·79	2·21	2·64	3·06	3·49
25	0·63	0·94	1·28	1·67	2·07	2·47	2·87	3·26
30	0·61	0·91	1·22	1·60	1·98	2·36	2·73	3·11

Income level 1 is under $4000
2 is $4000— 5999 per year
3 is $6000— 7999 per year
4 is $8000— 9999 per year
5 is $10 000—11 999 per year
6 is $12 000—14 999 per year
7 is $15 000—20 000 per year
8 is over $20 000 per year

Source: Thomas and Thompson (18)

effect of varying time values. In the event this made little difference to the comparison between the sites, since the differences came from the smaller number of journeys losing large amounts. A zero value for all time savings of less than 5 minutes would reduce the benefits of inland sites compared with Foulness by less than 1 per cent while a zero value for savings of less than 10 minutes would make a similar reduction of 2·5 per cent. Such a convenient outcome could not however arise, if the problem were one of determining the extra costs of a slightly more distant location within the chosen area.

The authors of the M.1. Study (19) recognized that there are indivisibilities with freight transport:

... not all the time a vehicle is in use is spent running: some is spent loading and unloading, cleaning etc. Reynolds and Beesley have given information which indicates that about three-quarters of the utilised time of road vehicles gainfully employed (i.e. excluding cars in non-working time) is journey time which is susceptible to effective savings in time.

Surprisingly the M.1. Study completely ignored the existence of similar indivisibilities with business travel. The D.O.E. accepts the linearity assumption, although further doubt is cast on its validity by the priority evaluator study. This indicated that travellers attached relatively little weight to small time savings. Empirical work on small time savings has been made particularly difficult by the polarization of reported time differences/savings towards standard units such as 5 minutes.

5.4. DISCUSSION OF CURRENT PROBLEMS

5.4.1. JOURNEY PURPOSE

The purpose of a journey may determine the value of any time saving. People who are travelling during the course of their business day might be expected to use any time savings for work purposes. If 'unproductive' travelling can be used for work then the effective working day is extended. The best available estimates of the value of the extra output obtained from the longer effective working days are gross hourly labour costs. Average earnings are therefore used to evaluate business time savings. This averaging process implies that all time savings can be used productively (the problem of small time savings is discussed in 5.4.5. below) and that travel time is necessarily unproductive. The second assumption is not always valid for journeys made by public transport. Businessmen commonly work on inter-city rail journeys and it may be that it is inappropriate to use average earnings figures to evaluate time savings on these routes. As is the case elsewhere, the validity of one averaging process partly depends upon what other averaging has been carried out. Thus if the time saving values on an inter-city train are based partly on the high average earnings of the type of businessman who uses them, some adjustment to allow for the productive use of the travelling time should be made. On the other hand, if the 'business' part of the train time savings is valued according to average national earnings, then this may offset any over-estimate of time savings benefit caused by ignoring the possibility that travelling time can be used productively. The use of the same time values for all business travellers also fails to take account of a possible difference between the positions of self-employed workers and employees. Self-employed workers may have flexible working hours and self-imposed 'work-loads' which they will try to perform each day. If their work involves travel the working day may be extended to cover travelling time. Any travel time saving will not then add to work output but will increase available leisure time. Some top level executives who earn high annual salaries, but are not eligible for overtime payments, may be in the same position as self-employed people who use travel time savings for leisure rather than work activities. The use of their average earnings figures may therefore over-estimate the value of travel time savings to these categories of people. On the other hand average leisure values may under-estimate the time savings values, since the marginal value of leisure might be high for those with an extended working and travelling day.

Although this is not strictly part of the averaging problem, it is relevant to point out that the use of earnings figures for evaluating travelling time savings does not allow for the utility or disutility of travel to the business traveller himself. Some business travellers may prefer travelling to being at work (so that travel time savings which must be spent at work will have a negative value for them) while others will prefer work to travel. This may explain the low values for working time found in the Edinburgh/Glasgow study (11).

In Britain all non-business travellers are grouped into one category and average time values are used for the whole group. (In France even the distinction between business and leisure travel is not made and the same average time values are used for travellers in both categories.) The group who are not earning any income while they are travelling is clearly very large and must include people who would put widely different values on time savings. The young lover and the Sunday pleasure motorist have already been mentioned; commuters, shoppers, housewives, old-age pensioners, those visiting sick relatives, people travelling to take part in sport or to be entertained, and children being conveyed to or from school all belong to this large, heterogeneous class of 'leisure' travellers.

While it is apparent that some averaging of the time values of leisure travellers is necessary, it is not obvious that a single average value is adequate for all forms of road investment appraisal. Most work dealing with the value of leisure time has focused on measuring these values, and little attention has been given to developing any theoretical basis for them. Current empirical evidence in Britain suggests that leisure time should be valued at about 25 per cent of the average wage rate, or at 19 per cent of average household income. Some measurements have, however, shown much higher values (See Table 5.1) and the 25 per cent figure cannot be said to rest on completely unshakable foundations. Valuation of all leisure travel time savings at 25 per cent of earnings does not fit in with conclusion based on theoretical considerations. Although workers may not be entirely free to exchange leisure for paid work in the way assumed in some theoretical discussions of wage rate determination and the supply of labour, there should be some coincidence between the marginal value of leisure and earnings levels. Workers can affect the length of their working day through collective bargaining and, individually, by deciding whether or not to work overtime or to take on extra paid work. Workers usually demand a marginal wage rate that is higher than the average hourly rate if they are to exchange extra leisure time for work. This suggests that marginal additions to leisure should be valued at the same rate as earnings. But there are some difficulties in applying this argument to travel time savings. The disutility of spending time travelling might be much less than the disutility of being at work, though there are no reasons why this should necessarily be so. There is also a difference between what people are prepared to pay to enjoy travel time savings and what they will demand as 'compensation' from their employers for giving up leisure time. These two factors could explain the low 25 per cent of earnings valuation of travel time savings, though the explanation appears to be extremely tentative.

Another possibility, directly relevant to our discussion of the averaging problems, is that travel time savings values for one important class of journey, the journey to work, may be higher than those for other forms of leisure travel. Leisure time may be regarded as time which people are free to allocate to any purpose they choose although, as Evans and others have pointed out, some activities demand the consumption of indivisible 'lumps' of time. The principle of diminishing marginal utility appears to apply to increments of leisure time. Other things being equal, people who have an abundant supply of leisure time may be expected to attach lower values to small increments of it than those whose initial stock is small. This, of course, is implied by the theory linking the marginal wage rate to the marginal value of leisure. But a man's stock of leisure time must be related to his 'day', or to the interval between successive periods of sleep. Suppose that a man normally spends 8 hours sleeping so that he has 16 hours of waking time per day. On a working day he may spend 8 hours at work and one hour travelling. If another 2 hours are used up in essential activities such as eating, dressing, and domestic chores only 5 hours of freely allocable leisure time would be available. A recent study of the daily activities of 30 000 people in 12 countries found that physical and socio-economic needs took from four-fifths to five-sixths of the total time available on a working day (23). On Saturday and Sunday, however, 14 hours of free leisure time may be available. Since time savings on a Sunday cannot be transferred for use on a weekday, an increment of 15 minutes on a working day would be of more value than a similar increment at the weekend. The principal possible source for time saving on a working day is likely to be the journey to and from work. It can therefore be argued on theoretical grounds that there is a case for attaching a higher value to saving in commuting time than to other leisure travel time savings. The argument could also be applied to time savings on working days other than those on the journey to work. The Edinburgh/Glasgow data (7, 11) appears to contradict this, but the analysis may be affected by the exclusion of an income variable. People making inter-city shopping or recreational trips on working days may have higher average incomes than commuters. In the necessary compromise between using average and individual time saving values it must probably be accepted that it would not be practicable to use different 'pure leisure' time saving values for different days of the week. But commuter travel is such an important factor on many roads that separate time values should be used if it is indeed true that average commuter travel time savings have a higher value than average pure leisure time savings.

The index of total weekly hours worked by all operatives in manufacturing industries (1962 = 100) fell from 104·6 in 1956 to 81·1 in 1972 (24). Where a fall in total hours worked per week represents a change from a 5½ to a 5 day week the value of commuting travel time saving relative to weekend travel time savings would be increased. The reverse would be the case where the length of the average working day is shortened. The empirical evidence does not give a consistent picture of the relationship between 'pure' leisure and commuting time saving values. The autostrada study, for example, showed that commuting time savings were more highly valued than 'other non-work' travel time savings for travellers in large cars, but the reverse was found for occupants of medium-sized cars (15).

5.4.2. INCOME

Income levels are likely to have some influence on the price people will be willing to pay to gain travel time savings, although the empirical evidence (see Table 5.2) is not wholly consistent. Where price differences are small, as with the Sydney Ferry Study (17), non-income factors may determine modal choice. But the probability of choosing a quicker and dearer, rather than a slower and cheaper mode of transport is likely to increase with income level. It would be very surprising, for example, if the average income of passengers using long-distance coach services in Britain was not lower than that of people who travel by rail on inter-city routes.

The extent to which account should be taken of the effect of income on travel time savings values must ultimately involve making value judgements. The arguments relating to working time may be rather different from those applied to leisure time savings. If earnings are related, even in a somewhat loose and erratic manner, to the value of work output then there is an obvious case for basing work time saving values on earnings. While the community accepts different earnings levels for different jobs it is logical to apply the same differentials to work time savings. For leisure time savings the 'equity' argument is that the influence of income should be ignored and that the same unit values should be used for all leisure time. The community will not gain when an individual's leisure time is extended in the way that it would from a lengthening of his effective working day. There is an obvious attraction in the value judgement that in the evaluation of leisure time all men should be treated as equal, especially when the wealth enabling richer people to put a high value on leisure time savings comes from unearned income. On the other hand it can be argued that higher-income earners should be allowed to buy extra leisure in the same way that they can buy better houses or more luxurious cars. High earned incomes may be the result of working for long hours, or at a very intensive rate, so that additional leisure may be highly valued because it is especially needed.

The current philosophy of the D.O.E. is to recognize the importance of the income factor in valuing working time savings (see Table 5.1), but to recommend a standard 'equity' value for all non-working time savings when these are used for investment appraisal. For behavioural studies which seek to explain the way in which travellers choose between different modes of travel, and to forecast traffic flows and modal split, separate leisure and commuter time values are suggested which vary with income levels. Thus the 1968 average commuting time values for car drivers was assumed to be 18·0p per hour, while that for bus passengers was estimated to be only 10·25p per hour (25).

In evaluations of road investments the significance of the effect of income (and of all the other factors discussed below) on time values may be largely removed through the use of national average figures for traffic composition and time values. The Road Research Technical Paper 75 (26) has been widely used by local authorities for the appraisal of road projects. In this paper a comprehensive formula:

$$C_m = 4\cdot3 + \frac{230}{V} \text{ (old pence) } (V < 38 \text{ mile/h})$$

is given for the value of all time savings (both personal time savings and vehicle savings). To produce this formula national average time saving values for each of five classes of vehicle were estimated. Personal time savings values were based on estimated average wage rates for working time and on 75 per cent of average wage rates for leisure time for each class of vehicle. A weighted average of the values for the five vehicle classes was then calculated using figures for the average composition of traffic on all roads in 1964 as weights. Where the 'Dawson formula' is used the income of travellers will have no direct influence on the ranking of alternative road investment projects. Roads with a high proportion of high value time-saving travellers (businessmen at work, car drivers, and passengers), or in high-income areas, will not gain any corresponding increase in the value of gross benefits in an investment appraisal. The level of gross benefits for any given speed increase, when calculated from the Dawson formula, will be determined only by traffic-flow.

The D.O.E. now recommends the use of the 'cost benefit' appraisal technique known as COBA rather than R.R.75. The COBA programme does not allow local authorities to feed in local values for working time, but the D.O.E. encourages the use of local data on traffic composition. In both cases national average figures are otherwise used.

A recent T.R.R.L. publication (27) gives details of a model, called CRISTAL, which can be used to plan a strategy for urban transport investment. In CRISTAL higher time values (30p per passenger hour).

are suggested for buses than for car travellers (20p per passenger hour). The reasons for this (which are discussed more fully in section 5.4.3.) are that the effect of income is excluded and account is taken of peoples' relative dislike for bus and rail travel. The CRISTAL time values make no distinction between working and non-working journeys (except for bus and lorry crews and taxi drivers) or between commuters and 'pure' leisure travellers. In practice there are also differences in dealing with the income factor in investment appraisal in other transport modes. The Roskill Commission used high time values, based on the average earnings of businessmen travelling by air, of 231·25p per hour for working time in evaluating the cost of travelling to alternative sites for the proposed third London airport. This value was much criticized, as the cost of travelling to the site was the main factor which suggested that the airport should be located at Cublington. Details of rail investment appraisal procedures are not revealed. Investment in rail electrification has been justified by British Railways, ex post, because it has increased passenger traffic and revenue. The revenue received from selling seats in a train must be related to the 'willingness to pay' (although it does not measure consumer surplus) of rail travellers which may reflect their relative affluence.

The policy implications of decisions about how far the income factor is allowed to influence the evaluation of time savings are not entirely straightforward. The use of income-influenced time values would of course favour investment in roads carrying a high proportion of travellers who put high money values on time savings. The time savings of groups of people with higher average incomes (such as car drivers) would become more important per minute than those people in lower income groups (such as bus passengers).

Investment in parts of the country where average income levels are higher (such as the South East) would also be favoured compared to investment in lower average income areas (such as the North East). When only buses and cars are considered, it is possible to estimate bus/car ratios, showing the number of cars which yield the same value of personal time savings per hour as one bus. Suppose the occupancy figures and proportions of working time are the same as those given in R.R. 75 (car 1·4 persons in working time, 2·1 in leisure time, bus 20 persons; car 21 per cent working time, bus 3 per cent), that bus personal working time is valued at 100p per person per hour, that car personal working time is valued at 200p, and that the 'equity' value of leisure time is 37·5p per person per hour (or 25 per cent of the unweighted average value of bus and car working times). The bus/car equivalents can be calculated, on the basis of these assumptions, for five different income situations. These are: 'current theory', where working time is valued according to average earnings, while a standard equity value is applied to leisure time; 'standard traffic', where the Dawson formula with standard traffic composition and time values is used; 'all equity', where all time savings are valued at the equity rate; 'all income', where working time values are based on earnings, while leisure time values are set at 25 per cent of the average earnings for each vehicle class; and CRISTAL, where the 'disutility factor' is assumed to be more important than the income factor. The bus/car ratios are then:

current theory	1 : 6·5
standard traffic	1 : 1
all equity	1 10·2
all income	1 : 3·8
CRISTAL	1 : 15·4

Another way of looking at the policy implications of the above analysis is to compare two roads with slightly different bus/car ratios. Suppose the average hourly traffic flows were:

| Road A | 28 buses | 900 cars |
| Road B | 27 buses | 904 cars |

and that all other factors (costs of road improvement, vehicle time savings, accident savings, and changes in maintenance costs) were the same. Then the value of personal time savings would determine which of the two roads (assuming they were exclusive) should be improved. These values (using the same assumptions as for the ratio calculations) are shown for each of the five income situations in Table 5.5.

TABLE 5.5   *Daily value of benefits from a one-hour reduction in average journey times, Roads A and B*

Income treated in terms of:	Gross value of personal time saving benefits	
	Road A	Road B
	£	£
Current theory	1309·6	1306·6
Standard traffic	1246·7	1250·7
All equity	869·1	864·6
All income	1428·3	1428·6
CRISTAL	519·5	515·1

Three of the income factors would thus show that Road A was the preferred investment, whilst the other two show improvements to Road B yielding the greatest benefit. It is particularly interesting to note that the 'standard traffic' approach of R.R. 75, although removing the direct influence of income on time saving values, is likely to work in favour of roads with a relatively high proportion of car traffic. It thus has the same general effect as using time values determined by income levels.

There is an interesting contrast between the treatment of income level differences in transport investment appraisal and the arguments for imposing a congestion tax on vehicles using congested urban roads. In the latter case both the original Road Pricing White Paper (28) and recent T.R.R.L. publications (29) assume that the pricing mechanism should be used to allocate road space and that it is not necessary to introduce weights to correct for any unacceptable features in the distribution of income. Failure to establish an agreed relationship between income and the value of time savings means that forecasting future time saving values on the basis of assumptions about changes in income levels is not entirely satisfactory.

5.4.3. TRANSPORT MODE

Both theoretical considerations and empirical evidence suggest that some forms of travelling are regarded as less pleasant than others. This means that, *ceteris paribus*, people will value time saved on a less pleasant transport mode more highly than savings on a preferred one. It is generally assumed that bus travel is less attractive than car travel (see for example, Quarmby (6)) so that for a typical traveller a five minute reduction in the time taken on a bus journey should be worth more than a similar reduction in a car trip (assuming the journeys were originally of approximately equal duration). The present discussion relates only to in-vehicle time. The advantage of the car is likely to be increased when terminal time (walking and waiting) which is more disliked than in-vehicle time is taken into consideration. T.R.R.L. figures used in operating the CRISTAL model, for example, show that average bus terminal times were 0·24 hours per person trip in Inner London in peak periods in 1970, while average car terminal times were 0·06 hours per person trip (30).

The above relative travelling time values of the CRISTAL model, e.g. 20p for cars, 25p for rail and 30p for buses (per person per hour), were determined by the estimated differences in travel disutilities. The initial assumptions were of a single value for all passenger travelling times of 25p per hour (and 50p per hour for terminal time), but it was decided that some preference for car travel and some dislike of bus travel ought to be displayed in the values of time used. The above arbitrary adjustments were therefore adopted (27). It is difficult to reconcile this approach with the use of income-influenced figures by the D.O.E. in its instructions to local authorities on road investment appraisal. The higher income levels of average car travellers may well offset the lower disutility factor in determining the net relative time savings values for car and bus travellers. Suppose that average hourly earnings are represented by W_0 and W_1 (for car and bus travellers respectively); the proportions travelling in the course of work are P_0 and P_1; the relative disutility factors are U_0 and U_1; and leisure time is, valued at 25 per cent of working time. Then the value of car travellers' time (in pence per person per hour) will be:

Car time savings value $= P_0 W_0 + (1 - P_0)(\cdot 25 W_0) + U_0$

$$= 0\cdot 75 P_0 W_0 + 0\cdot 25 W_0 + U_0$$

If the 1972 D.O.E. time-saving values of 96·0p and 183·5p are used
for bus and car travellers, and it is assumed that 10 per cent of
travellers in both modes are making work journeys (average figures
in the 1964 National Travel Survey showed that 7·9 per cent of travel
mileage was 'in course of work'), then car and bus time savings would
be valued as follows:

  Car time savings $= 59\cdot 6 + U_0$ pence per person per hour
  Bus time savings $= 31\cdot 2 + U_1$ pence per person per hour

This means that the disutility factor would need to be more than
28·4 pence per traveller per hour before bus traveller time savings
had a higher unit value than car traveller time savings. This is a very
high value and it seems most likely that time savings per person should
be valued more highly for car than for bus travellers unless the income
factor (and most empirical 'willingness to pay' evidence) is discounted.

There are, however, further problems about the evaluation of
different time-saving values which are illustrated by the CRISTAL
figures. The CRISTAL time values are intended for application to a
model which could be used to compare alternative urban transport
strategies rather than for evaluating road investment projects. If
applied to the latter, CRISTAL time values would result in higher
rates of return on investment in those roads carrying a high proportion of bus traffic. But when used to devise an over-all transport
strategy, where it is possible to choose between the development of
alternative modes, the high bus time values become a 'cost' and dis-
courage rather than support investment in buses. When evaluating
those road investments which attempt to increase traffic speeds, it is
permissible to add the effects of the varying income levels associated
with different transport modes, together with the influence of the
'travel disutility' factor. If one mode was both most disliked and used
by people whose high incomes led to correspondingly high time
savings values, then both factors would support investment in the
roads on which that mode predominated. As might be expected, the
evidence in facr suggests that higher-level income groups use the
transport mode which is less disliked and it seems likely that in general
(though there are exceptions) the influence of income outweighs
that of the differences in disutility. The use of a single time 'cost'
in the CRISTAL model is potentially confusing. If it is planned to
carry a particular group of travellers, either all by bus or all by car,
then the disutility factor is the only one which need be considered.
Personal differences in valuing time savings (which may be largely
determined by variations in income level) will not matter if only a

single mode is involved. But in any case, where both modes are used, differences in personal time-saving evaluations may be associated with the mode that is relevant. A bus-lane scheme, which would result in delays to cars, might be shown to be beneficial if bus time savings were valued more highly than car time savings (to reflect differences in disutility). It might nevertheless be shown to yield a net disbenefit, when time savings are valued on a 'willingness to pay' basis reflecting differences in income levels as well as relative disutilities.

As already argued in 5.4.2., the ranking of different road investment projects will not be affected by the level of the personal time savings values attached to different travel modes when average traffic composition figures are used. If actual local traffic flow data were used, the rates of return on urban road improvements would be raised relative to those on rural roads (because of the high time-saving values given to bus travellers).

5.4.4. TYPE OF TRAFFIC

A major distinction in traffic types is that between goods and passenger traffic. The same general principles are applied in Britain in evaluating the time savings of both goods and passenger traffic. In R.R. 75 three main components of vehicle costs which vary with journey time were distinguished. These were 'labour costs' (equivalent to personal time-saving values), overhead costs and speed-dependent operating costs. The values used for cars, buses and heavy commercial vehicles are shown in Table 5.6 (in old pence).

TABLE 5.6   *Time saving values for buses, cars and lorries, R. R. Tech. Paper 75*

Vehicle	Personal time savings	Overhead costs	Speed-dependent operating costs	Total
Bus	1090	67	31	1188
Car	203	4	25	232
Heavy lorry	90	32	23	145

These figures, which suggest that the value of a given time saving for a heavy goods vehicle is less than that for a car or a bus, require some explanation. The overhead cost item is not explained in R.R. 75, but presumably reflects the benefits from better vehicle utilization which would result if a bus or lorry could perform a greater annual mileage as a result of increased average journey speeds. The main item would be vehicle depreciation which is largely determined by the cost of the vehicle. But overheads appear to be higher than those for lorries because the 'heavy' lorry value used was for a 'typical' vehicle of only 8 tons carrying capacity, and this would cost considerably less than the 44 seater bus used to represent public service vehicles. The speed-dependent operating costs were lower for lorries than for cars, mainly because reductions in fuel costs (for increases up to 40 m.p.h.) were estimated to be less for lorries than for cars (or buses). But the factor which determines the relative vehicle time saving values, more than off-setting the car/lorry difference in hourly overhead costs, is personal time savings. Because car travellers have higher average incomes than lorry drivers, and because average occupancy is higher for cars than for lorries, 'per-vehicle' personal time savings in cars were valued at more than twice the level for 'heavy' lorries. The car personal time-saving values were particularly high in R.R. 75, because leisure time values were assumed to be 75 per cent of the 'relevant' average wage rates. If this is replaced, as current D.O.E. theory suggests, by leisure time valued at 25 per cent of national average wage levels this would clearly reduce the value of car time savings considerably.

If 1974 D.O.E. personal time saving values and R.R. 75 occupancy levels and leisure/work percentages are used, comparisons of the values of personal time savings per vehicle can be made for cars, 'heavy' lorries, and buses. For a car travelling in working time the values would be:

Car = 331·8p per vehicle hour
Lorry = 137·7p per vehicle hour.

Personal time savings for an 'average' car (with 21 per cent working time and 79 per cent leisure time), and with leisure time valued at 25 per cent of the average of car and lorry working time, would be:

'Average' car = 142·6p per vehicle hour.

If the updated leisure value of M.A.U. 179 (25) were used, then the figure would be:

'Average' car = 115·0p per vehicle hour.

Bus personal time saving values, with leisure time as 25 per cent of the average of bus and lorry working time, would be:

Bus = 868·6p per vehicle hour,

or, using M.A.U. 179 leisure values, they would be:

Bus = 801·0p per vehicle hour.

In this case the addition of bus overheads and speed-dependent operating costs (as calculated in R.R. 75) would increase the gap between bus and lorry time saving values.

If the allocation of investment funds to roads were decided entirely on the basis of an economic assessment, using 'official' time values (and local traffic flow data), then roads carrying a large volume of goods traffic would have a relatively low priority. One heavy lorry would only have approximately the same importance as one car and less than one-fifth that of a bus. But there are many more cars than lorries on the roads. Even on motorways, which carry a relatively high proportion of goods traffic, typical figures for 24 hour flows at selected points in 1971 were (31):

M.1	Cars and taxis	=	53·32 thousand vehicles
	Goods vehicles (except light vans)	=	11·90 thousand vehicles
	Ratio	=	3:1
M.6	Cars and taxis	=	21·38
	Goods vehicles	=	9·53
	Ratio	=	2:1

The flow of cars would thus be the main factor determining road investment. Roads carrying holiday and pleasure traffic (even after allowing for the peaked nature of demand) could thus take precedence over important lorry routes. If this is unlikely to happen in practice, it is because an economic evaluation does not always determine final investment decisions (or because lorry and car flows may be positively correlated on some roads). Government policy statements have often suggested that the need for good routes for transporting industrial freight was an important factor in determining road investment. Thus the 1972–3 report, Roads in England, states that 'the objective of linking major industrial centres with the ports is exemplified by the continued work on the Airley Top to Chain Bar section of the M62' (32). It is obviously unsatisfactory if the economic appraisal of road investment projects does not measure all the benefits which result from reducing the journey times of goods vehicles.

The basic inconsistency in evaluating the time savings of goods vehicles is that no account is taken of the benefits of reducing the journey time of the consignments carried. Neither the driver's wages nor vehicle overheads is an adequate proxy for what might be called the 'load' time savings. There is ample evidence to show that consignors of freight transport are often prepared to pay substantially higher haulage charges in order to achieve quicker deliveries (33, 34). Reductions in journey times may be particularly valuable for high cost consignments; for export traffic being carried to a port; for repaired or replacement machinery and spare parts; and for perishable goods. The most general benefit is through reductions in the cost of stock holding. This item alone would amount to about 17·0p an hour for a consignment worth £10 000 (assuming a 15 per cent annual interest rate). Bus vehicle time savings values are strongly influenced by occupancy rates and if average bus occupancies increased, time saving values would need to be adjusted upwards. But an increase in average lorry loads would not affect lorry time saving values (except to the very limited extent to which average drivers' earnings might be raised if more high capacity lorries were in use). This seems to be an anomaly. In the Channel Tunnel study a four-hour time saving for lorries is estimated to be worth only £6 (equivalent to the saving in drivers' wages) (14). It is difficult to believe that this measures the full extent of the benefit from such a large journey-time saving. There seems to be a strong case for adding an allowance for the benefits from delivering consignments more quickly. The present practice of valuing an hour's time saving for an average heavy lorry at about the same level as an hour's time saving for an average car is hard to justify.

5.4.5 SIZE OF TIME SAVING

Current practice in transport investment appraisal involves giving the same average unit value to travel time savings, irrespective of the size of the saving. If time savings are valued at $£x$ an hour, then their value is also assumed to be $£\frac{x}{60}$ per minute and $£\frac{x}{3600}$ per second. The major exception to this was the report on the Third London Airport in which, although constant unit time values were employed in all the main calculations, the effects of giving a zero value to savings of less than 10 minutes, and of less than 5 minutes, were also estimated (13).

The argument that there are indivisibilities in the use of time, and that time savings which are too small to be used have a low or zero value, has been put forward by Tipping, Mishan, and others (35, 36). It is obvious that there are individual cases where small time savings have little or no value. A business executive may not be able to do any useful work at all in an isolated period of a few minutes. Similarly, a man may value an extra hour at home, where he can tackle the garden or undertake some extra leisure pursuit, whereas an additional two minutes may not have any use. The basic argument for giving full unit value to small time savings is that successive small increments of time savings may add up to a usable 'lump' of time. Suppose that an improvement to the road between Birmingham and Leicester reduces the journey time by 3 minutes, but that businessmen travelling on this route can only use time increments productively if they total at least 30 minutes. It may be that previous improvements have already given time savings of 27 minutes on this route, and that the extra 3 minutes will turn all these 'unusable' time savings into usable increments. There is certainly some force in this argument. It would be absurd to give a low or zero value to a succession of minor improvements on a route used by through traffic when it was clear that the total time saving over the route from all the improvements was substantial. The time savings on each link road and intersection of an urban network road improvement scheme may be valued separately, but most traffic will probably use a large part of the network and thus enjoy relatively large total time savings. But it is not necessarily true that all small time savings can be aggregated to form usable time increments. Many urban commuter routes, on which the great bulk of traffic makes quite short journeys, may only have one or two improvements, yielding very small individual time savings over a number of years. There may be no way in which these time savings can be accumulated to form a usable increment of time. The commuter will not be able to add together separate time savings of 3 minutes a day to give him an extra hour in the garden every 20 days. Time savings cannot be stored, but must be consumed as they occur. A commuter cannot even add together time savings made on outward and return journeys (unless his employer allows him to arrive at work early and to leave for home early).

There thus seems to be a strong case for arguing that for at least one class of small time savings, e.g. those occurring on road (or rail) routes used predominantly by relatively short-distance commuters, lower unit values should be used. However, even if time savings cannot be put to any use, this does not mean that they have zero value. The 'disutility' factor, as well as the opportunity cost, must be taken into consideration. People may gain from having shorter journeys, even if they make no use at all of the time saved, because the unpleasantness associated with the journey itself is reduced. It can also be argued that the minimum period of usable time may be much shorter for leisure than for business time. Another two minutes reading the paper at home may be valued; an extra two minutes inside an office or factory may not yield any increase in output. Following the argument of 5.4.1. it may be that small time savings have a greater value on workdays than on non-work days. If small transport time savings which cannot be aggregated were given a lower unit value than large time savings, this could have important policy implications. It would mean that the rates of return on urban network improvements might be over-estimated in relation to those of long sections of trunk road or motorway. Investment on a small scale in any form of transport (where time saving values rather than money returns are used to measure the expected benefits) might be unduly favoured compared with large-scale projects. With given budget constraints there might be a choice in a conurbation between building a number of new road underpasses on different commuter routes into the central business area, or constructing a single extension to an underground rail system. If the underpass showed a slightly higher rate of return when constant unit time values were used it is very probable that devaluing time savings of less than about five minutes would decisively reverse the relative ranking of the two projects. The only empirical evidence available supports the suggestion that small time savings have a lower unit value than larger ones.

5.5. THE GRANBY HALLS SCHEME

Some of the implications of disaggregating time saving values can be illustrated by examining an actual urban road investment appraisal.

The Granby Halls project in Leicester, details of which were kindly
supplied to us by Mr. Derek Sharpe, now the City Engineer, is a typical
medium-sized road improvement scheme. As originally designed, the
scheme involved turning an existing road into a dual carriageway, the
building of two short stretches of new road (total length 0·31 miles),
and the alteration of the traffic flow system over a number of streets
about half a mile from Leicester city centre. The estimated cost of the
original scheme was £1·39m. The base date for the calculations was
1969 and an estimate of the return was made for 1974. This rate was
estimated to be 20 per cent on the official calculations, though our
reworking of the data suggests that there was a small arithmetical error
and that, using all the 'official' assumptions, the 1974 first year rate of
return was 19·7 per cent. The calculations followed the procedure
suggested in R.R. 75 outlines in 5.4.2. above. The factors taken into
consideration were 'cost of delay', changes in accident rates, and changes
in road maintenance costs. Differences in the total route length before
and after the improvements were allowed for by using a delay cost
formula which included a running cost item varying with route mileage.
The single Dawson formula was used for all types of traffic, using a
weighted average of the values for five classes of vehicle—cars, light,
medium, and heavy goods vehicles, and buses. This weighting was based
on the average traffic composition of all roads in 1964 which was:

Cars	71·7 per cent
Light commercial vehicles	11·8 per cent
Medium & heavy goods vehicles	13·6 per cent
Public service vehicles	2·9 per cent

The Granby Halls scheme thus took no account of the actual composition of the traffic flows on the roads which would be affected by the
scheme. Total traffic flows were estimated, but it was assumed that the
traffic had the same composition, in both the type of vehicle and in
the proportion of business and leisure car travellers, as the national
average figures used in the formula. It is also assumed in the formula (and
thus also by those carrying out road investment appraisals) that the same
time saving values should be attached to commuter traffic as to all other
non-business traffic; that income factors should be ignored; and that
the time savings of goods vehicles is adequately represented by the
average value of the wages paid to the driver (or crew) plus some saving
in overhead costs and speed-dependent running costs.

The appraisal of road investment schemes like the Granby Halls
project could be made more sophisticated by using local rather than
national average traffic composition data; by taking account of more
variables in estimating the value of travel time savings; or by doing both.
There are two major results which might follow such a reappraisal. If
road transport investment projects are compared directly with alter-

native non-transport projects (or with some non-road transport projects), then a revised rate of return from the road investment might make it appear either more or less desirable. But the ranking of one road investment project relative to similar projects would not necessarily be changed, because all other comparable rates of return might be changed by the same proportion. Thus so long as a standard traffic composition is assumed, changes in the value of time savings of different components of the traffic flow will have no effect on the relative ordering of alternative projects. Unless rates of return in road investment projects are to be compared with those in non-transport investments, the use of standard traffic composition figures makes the whole business of trying to evaluate time savings a somewhat pointless exercise.

Details of the composition of the Granby Halls traffic were not available, but for the purpose of this exercies the results of two 'guesses', based on observation of the local traffic flows, were applied. These guesses or assumptions were that the Granby Hall traffic flow contained a higher proportion of buses than the national average figure of 2·9 per cent and that the proportion of commuter traffic would also be above the national average. Since commuter traffic was classed as non-business traffic, this means that the proportion of car non-working traffic may have been higher than the 79 per cent assumed in the Dawson formula. (The proportion of 'pure leisure' traffic may be low, but it is arbitrarily assumed that the proportion of commuter and 'pure leisure' traffic taken together came to more than 79 per cent). If these assumptions were correct and local traffic flow data was used to weight the time values which are combined into the Dawson formula, then each assumption would have an opposite effect on the rate of return. A higher proportion of buses would increase the rate of return, since buses are given the highest time value rating per vehicle mile of any class of vehicle identified in the R.R. 75 analysis. (The rate of return would be raised still further if the average bus occupancy were higher than the 20·0 passengers assumed in the Dawson formula.) An above-average proportion of non-working travellers in the traffic flow would reduce the estimated rate of return from the Granby Halls scheme.

As had been shown above, changing the time saving values of different types of traffic only has significance when a standard traffic composition is assumed. The influence of time value changes on rates of returns is obvious. Increases in component time saving values would increase the over-all rate of return. It has already been argued that there is a case for believing that the values for commuters and heavy lorry traffic should be increased, and the recommended value for leisure time savings has already been reduced from 75 per cent to 25 per cent of working-time values. If actual local traffic flow composition data were used and the suggested changes in time savings values were made, then the effect on the Granby

Hall rate of return could be estimated. The increased commuter time value would increase the Granby Hall rate of return relative to projects which had the same proportion of commuter traffic as the national average. The increase in lorry time value and the decrease in 'pure' leisure time values would have no effect on the ranking of the Granby Halls project because the local proportion of these traffics has been assumed to be the same as the national average.

It has also been suggested that small time savings might be devalued and that, if income levels are reflected in the relative time values of different forms of traffic, then it might be logical to allow for their influence on time savings values in different areas or regions. Since details of estimated traffic flows (but not traffic composition) before and after the improvement, and of estimated time savings on each link road and intersection, were available for the Granby Hall scheme it was possible to rework the Granby Hall calculations to discover the effect of devaluing small time savings. The results of devaluing different levels of small time savings are shown in Table 5.7. In each case the standard Dawson formula time value used in making the Granby Halls evaluation was devalued to 25 per cent of the full value. The only justification for the 25 per cent figure is that relatively small time savings may retain value for 'pure leisure' purposes so that leisure time values may be

TABLE 5.7 *The effect of different assumptions about the value of small journey-time changes on the estimated rate of return from the Granby Halls (Leicester) road improvement scheme*

Length of time change (mins)	Assumed time value p per hour	First year rate of return %
All	95·83	19·7
0·49 or less	23·96	
0·50 or more	95·83	21·8
0·99 or less	23·96	
1·00 or more	95·83	23·4
1·49 or less	23·96	19·9
1·50 or more	95·83	
1·99 or less	23·96	
2·00 or more	95·83	12·8
2·00 or less	23·96	
2·01 or more	95·83	5·3
3·00 or less	23·96	
3·01 or more	95·83	5·3

appropriate.

The Dawson formula time value was fairly close to the full average business time value, as the 'leisure' proportion of car and bus travel was valued at 75 per cent of working time. The argument for devaluing small time savings applies to personal time savings rather than to those vehicle running costs (or overheads) which vary with journey speed. Goods vehicles and buses are likely to be able to cumulate a series of small time savings in one day, so that vehicle utilization may be improved and daily overhead costs reduced. This is unlikely to be possible for cars used for the journey to work. It would, however, have been extremely difficult to disentangle the relatively small element of vehicle overhead and running cost time saving values in the Granby Hall calculations and the arbitary element in the 25 per cent figure would have made it some-what pointless to have tried to deal with this minor inaccuracy.

It may appear surprising that devaluing time savings of less than 1·49 minutes increased the estimated rate of return from this project. This was because some of the journey-time savings on intersections and links on the network of urban roads affected by the Granby Halls scheme were negative, i.e. journey times were increased. Since most of these increases were small, devaluing small time savings up to 1·49 minutes increased the net estimated rates of return. The large drop in the rate of return when time savings of 2·00 minutes were devalued was caused by a cluster of time savings at intersections estimated to be exactly 2·00 minutes. There were no links or intersection savings of 3 minutes, but this figure was included in the table to allow for the cumulation of time savings by vehicles using part of the improved road network. Clearly, if small time savings were given lower unit value than larger savings, then it would not be appropriate to evaluate the time change on each link and intersection separately. Allowance must then be made for the total effect on journey time for vehicles using the improved system. We calculated that (according to the local authority's figures) no vehicle would gain as much as 3 minutes in journey time by using the improved Granby Halls road system for any conceivable through journey.

5.6. SUMMARY OF POLICY CONCLUSIONS

(a) Whenever national average, rather than actual local, traffic com-position data is used in the economic appraisal of road investment projects misleading results may follow. The rate of return per unit of investment funds will be largely determined by the size of the total traffic flow (and that of the time saving per vehicle). The rate of return from a road investment project is generally compared only with that from other proposed road investments. (The reduction in the value of leisure time from 75 per cent to 25 per cent of working time has not had any apparent effect on the size of the road investment programme.)

The use of 'standard' traffic composition figures thus makes all attempts to find appropriate time saving values for different types of vehicle or traffic largely irrelevent. It also means that the benefits from any given reduction in journey time are treated as being of equal importance for one car, one bus, or one lorry. The first priority in improving techniques for evaluating transport time savings should therefore be to abandon the use of standard figures for traffic composition. This means that an aggregated formula of the Dawson type cannot be used.

(b) There are some quite strong theoretical reasons for supposing that commuting time savings may have a higher value than 'pure leisure' time savings. More empirical evidence is needed on this issue.

(c) Greater consistency is required in the treatment of the effect of income on the evaluation of personal time savings. There is a good case for treating leisure time and working time on a different basis, but the underlying value judgements need to be made clear and should be open to public discussion. A consistent national policy should be adopted on the use of local or national earnings figures for evaluating working time. If some local authorities use local figures while others rely on national data investment misallocations could result.

(d) There is a potential source of confusion in the treatment of the effect of people's preferences for different transport modes on the value of transport time savings. Where a transport investment will reduce the journey time of existing traffic flows then the alternative use value of the time saved, and the gains from reducing the disutility of travel, are additive. If people dislike bus travel more than they dislike car travel then this is a factor favouring investment which will reduce bus journey time. (Bus time saving values should, of course, be based only on in-vehicle values when an investment will not reduce walking or waiting times.) But when an over-all transport investment strategy (in which a choice between the development of different modes must be made) is under consideration then the situation is different. The degree of dislike for a mode represents a cost rather than a benefit, so that a greater dislike for bus travel is a reason for investing in rail or car transport. If the same people are to be transferred from one mode to another, then the income-influenced alternative use time value can be ignored.

(e) At present the value per unit of time saved is assumed to be about the same for a car as for a lorry, but to be much higher for a bus. The position of lorries appears to be anomalous and some account should be taken of the value of reducing delivery times for goods traffic.

(f) Where it is clear that travellers cannot aggregate small time savings of less than about 5 minutes then there is a good case for attaching a lower unit value to them. This may apply particularly to small non-work time savings at the weekend.

(g) The current practice of local authorities is to plan for an over-all transport investment strategy rather than to evaluate individual road projects. This makes it important to develop a clear and consistent policy towards the evaluation of transport time savings.

REFERENCES

1 Harrison, A. J. and Taylor, S. J., The Value of Working Time in the Appraisal of Transport Expenditures: A Review, *Research into the Value of Time*, D.O.E., Time Research Note 16, July 1970.

2 Becker, G. S., The Theory of the Allocation of Time, *Econ. J.*, 75, pp. 493–517, 1965.

3 Evans, A. W., On the Theory of the Valuation and Allocation of Time, *Scottish J. of Polit. Econ*, 19, pp. 1–17, 1973.

4 De Serpa, A. C., A Theory of the Economics of Time, *Econ. J.*, 81, pp. 828–46, 1971.

5 Beesley, M. E., The Value of Time Spent Travelling: Some New Evidence, *Economica*, 32, pp. 174–85, 1965.

6 Quarmby, D. A., Choice of Travel Mode for the Journey to Work: Some Findings, *J. Transp. Econ. Policy*, 1 (3), pp. 273–314, 1967.

7 Local Government Operational Research Unit, *Planning for the Work Journey*, Rpt. C67, 1970, *Modal Choice and the Value of Time*, Rpt. C143, 1973.

8 Social and Community Planning Research, The Importance and Values Commuters Attach to Time Savings, 1971 (unpublished).

9 University of Southampton, The Solent Crossing Study, 1970 (unpublished).

10 Veal, A. J., *The Value of Time on Short Urban Leisure Trips*, Centre for Urban and Regional Studies, University of Birmingham, 1970.

11 Watson, P. L., *The Value of Time and Behavioural Models of Modal Choice*, Ph.D. Thesis, University of Edinburgh, 1971.

12 Dawson, R. F. F. and Smith, N. D. S., Evaluating the Time of Private Motorists by Studying their Behaviour, Road Res. Lab., *Res. Note 3474*, 1959.

13 *Report of the Commission on the Third London Airport*, Roskill Report, H.M.S.O., 1971.

14 *The Channel Tunnel;* A Report by Coopers and Lybrand Associates Ltd., H.M.S.O., 1973.

15 Dawson, R. F., and Everall, P. F., *The Value of Motorists' Time: a Study in Italy*, T.R.R.L. Report LR 426, 1972.

16 Lee, N. and Dalvi, M. Q., Variations in the Value of Travel Time, *Manchester School*, 37, no. 3, pp. 213–36, 1969.

17 Merlin and Barbier, Private Commuter Travel Time Savings, Paris, 1962, cited in Henscher, D. A. and Hotchkiss, W. E., Choice of Mode and the Value of Travel Time Savings for the Journey to Work, *Econ. Rec.*, pp. 94–112, 1974.

18 Thomas, T. C. and Thompson, G. I., Value of Time Saved by Trip Purpose, *Highway Res. Rec.*, 314, Highway Res. Board, Washington, D.C.

19 Charlesworth, G. and Paisley, J. C., The Economic Assessment of Returns from Road Works, *Proceedings of Institution of Civil Engineers*, 14, pp. 229–54. 1959.

20 Foster, C. D. and Beesley, M. E., Estimating the Social Benefit of Constructing an Underground Railway in London, *J. Royal Statistical Soc.*, A (General), vol. 126, Part 1, 1963.
21 Information supplied by J. Stanley, Commonwealth Bureau of Roads, Australia.
22 Information supplied by A. Nichols, Department of the Environment.
23 Szalai, A. (ed.), *The Use of Time*, The Hague, 1972.
24 *Department of Employment Gazette*, Jan. 1974.
25 McIntosh, P. T. and Quarmby, D., *Generalised Costs, and the Estimation of Movement Costs and Benefits in Transport Planning*, D.O.E., M.A.U. Note 179, Dec. 1970.
26 Dawson, R. F. F., *The Economic Assessment of Road Improvement Schemes*, Road Res. Tech. Paper, no. 75, H.M.S.O. 1968.
27 Tanner, J., Gyenes, L., Lynam, D., Magee, S., and Tulpule, A., *Development and Calibration of the CRISTAL Transport Planning Model*, T.R.R.L. Report LR 574, 1973.
28 Ministry of Transport, *Road Pricing*, H.M.S.O., 1964.
29 Bamford, T. J. G. and Wigan, M. R., *The Effects on Transport Benefit Evaluation of User Misperception of costs*, T.R.R.L. Report 23 UC, 1974.
30 Holroyd, E. and Tanner, J., *A Simplified Form of the CRISTAL Transport Planning Model*, T.R.R.L. Supplementary Report 55 UC, 1974.
31 Department of the Environment, *Highway Statistics*, H.M.S.O., 1971.
32 Department of the Environment, *Roads in England 1972−73*, H.M.S.O. 1974.
33 Bayliss, B. and Edwards, S., *Industrial Demand for Transport*, Ministry of Transport, 1970.
34 Sharp, C., *The Allocation of Freight Transport−A Survey*, Ministry of Transport, 1970.
35 Mishan, E., What is wrong with Roskill?, *J. Transp. Econ. Policy*, Sept. 1970.
36 Tipping, D., Time Savings in Transport Studies, *Econ. J.*, Dec. 1968.

6

Resource Value of Business Air Travel Time*

by R. C. CARRUTHERS AND D. A. HENSHER

6.1. INTRODUCTION

Various aspects of the valuation of travel time savings arising from transport investment projects have received consideration in recent years. Initial attention concentrated on the journey to work as the journey purpose most likely to benefit from road and public transport investment projects. More recently attention has also been directed toward recreational and shopping travel, which has coincided with the re-orientation of government interest towards circumferential travel requirements in contrast to the previous emphasis on radial planning (1). Travel time savings on trips for business purposes have been relatively neglected, perhaps because of the small proportion of urban travel undertaken for either employers or personal business, but more likely because the underlying theory of the valuation of business travel time appeared to be relatively straightforward and to present no particular conceptual problems for valuation.

The basis of the traditional theoretical argument is that any travel time savings on business trips leads to an equivalent increase in available working time. This increased working time can be used to produce more output in the form of goods and services, or can reduce the cost of producing the same output. The net value of the increased output, or reduction in production costs, is measured by the average unit labour costs attributable to the type of employee travelling, i.e. the wage or salary costs plus any associated overhead expenses such as accommodation, equipment, payroll taxes, etc.

This argument is more easily accepted in the case of ordinary road travel for business purposes than it is in the case of air travel. For

* The study on which this chapter is based was commissioned by R. Travers Morgan and Partners in the course of a larger study they were carrying our for the Department of Transport of the Australian Government. Publication of this chapter does not imply acceptance or rejection of its conclusions by the Australian Government.

transport drivers and commercial travellers who make up the majority
of business road travellers a reduction in travelling time can quite easily
be seen, subject to scheduling and institutional constraints, to lead to an
equivalent increase in output or a reduction in the labour input required
to achieve the same output. But for business air passengers this is not
necessarily the case. For example, where the business trip is to attend
a day's meeting in another city, a reduction in travel time may only
result in the businessman spending the extra time at each end of the day
at home.

Some of the issues peculiar to business air travel time were raised in
the proposed Research Methodology of the Commission on the Third
London Airport (2). In Particular, it discussed the possibility that useful
work might be done while travelling and that savings in travel time
might lead to an increase in leisure rather than working time. Both
effects which, if allowed for, would tend to reduce the value of business
travel time savings (*VBTTS*), were dismissed on the grounds that they
were likely to significantly alter the *VBTTS*. No behavioural evidence
was produced to support this assertion.

The Roskill Research Team therefore ignored these issues, although
they were raised, together with a number of others, at the Commission's
public inquiry. These other issues centred on the method of determining
the unit value of time saved. First, it was claimed that a businessman's
day is not of equal productivity throughout, and that the rational
business traveller will organize his day so that only marginal activities
are given up for travel. Thus a saving in travel time will only increase
productivity at a marginal rather than at an average rate. Second,
doubts were raised concerning the standard method of dividing total
annual costs by the average hours worked in a year, to estimate the
average cost of an hour's labour. It was argued that businessmen are
employed to do a job rather than to work a fixed number of hours, and
that a decrease in travelling time would only reduce the time necessary
to perform set tasks, which would result in an increase in leisure time.
It was also argued that a businessman travelling and not working was not
contributing to his employer's profits, so that any reduction in travelling
time would lead to an increase in the amount of profit. The argument
that travel time is not a total loss to the business traveller was developed
further than in Roskill's Proposed Methodology. Whereas the Roskill
team had considered the possibility of activities such as reading while
travelling to be a form of constrained leisure which conferred no
measurable benefit, they claimed at the inquiry that significant amounts
of useful work could be, and in fact were, undertaken while travelling.

The principle of evaluation used by the Research Team ignored any
possible gain or loss to the traveller himself. It is possible that a businessman making frequent air journeys finds them less satisfying than working

at his office and therefore suffers some personal loss of utility.

Although all these possibilities were discussed at considerable length, there was little evidence, other than from personal experience or observation, to support them. The residual doubt in the minds of the Commission can be seen from the wide range of values adopted in their Report (from £3·05, or 94 per cent of the estimated hourly wage rate, to £6·20, or 167 per cent of the wage rate) and their stated sympathy with the view that the valuation of time is nothing more than an educated guess (3).

In 1972 the Australian Government, together with the State Government of New South Wales, commissioned R. Travers Morgan and Partners to investigate the proposed development of a Second Sydney Airport. At an early stage of the work it was realized that business travel time savings were likely to play an even more significant role than they had in the Roskill deliberations. Whereas at London Airports in 1968 32 per cent of passengers had been travelling on business, at Sydney in 1973 the figure was 46 per cent, made up of 23 per cent international and 52 per cent domestic passengers. An initial comparison of two potential new airport sites indicated that more than 20 per cent of the difference in net present value between them was attributable to savings in business travel time when valued in the simple Roskill way.

The study described in the remainder of this chapter was undertaken to explore many of the questions raised at the Roskill inquiry, and to derive a sounder basis for the valuation of business travel time savings. It aimed to establish resource values for evaluation, rather than behavioural values for predictive modelling, although a behavioural value was also obtained (4). The study also considered the incidence of the benefits of savings in travel time depending on whether the beneficiary was the business traveller, his employer, or the community as a whole.

6.2. THEORETICAL BASIS OF VALUATION

Our general theoretical approach was based on the traditional argument presented earlier, i.e. that the opportunity cost of business travel time is the value of the lost output or services that would otherwise have been produced. The main part of the study was designed to put a value on the ordinary working time of the business air passenger and relate this to the opportunity cost of travel time. We were particularly concerned to estimate the proportion of the business travel time savings that would produce an increase, not in productive output, but in the passenger's leisure time.

6.2.1 ISSUES CONSIDERED

Apart from determining the unit value of ordinary working time, the main issues which had to be considered were:

(1) the effect of some business travel being done in the employee's rather than the employer's time;

(2) the allowance to be made for work done while travelling; and

(3) whether travel time savings result in a saving of working time of average or marginal productivity.

The problems of trying to identify the marginal productivity of a business traveller's working day were daunting. Because the size of time savings might be considerable for some journeys, we would have needed to establish a marginal productivity function, which would have shown how the average *VBTTS* increased as the time savings became less marginal. It would also have been necessary to assume that the traveller was able to assess his own marginal productivity and, given more working and less travelling time, that he was capable of only increasing his productive activities. In the short term, the activities likely to be undertaken while travelling may have a relatively low productivity. In the longer term, on the other hand, it can be argued that a businessman's day cannot be disaggregated into more or less productive time—all activities contribute to an average productivity. As the values of time savings are for use in long-term investment programmes, long-term values are thus more appropriate than short-term ones. We did not consider these arguments further, and the study is incomplete in that the *VBTTS* derived is based on an average rather than a marginal cost function.

6.2.2. JOURNEY STAGES AND PASSENGER TYPES

The effect of travelling during the employee's own time, together with the fact that some work is done while travelling, was expected to be specific to each stage of a particular business journey, and to vary between domestic and international travel. (Domestic travel was defined to take place wholly within Australia.) For domestic travel we considered a total journey to comprise six stages:

(1) access to the airport on the outward trip;

(2) in-flight outward;

(3) egress from the airport outward;

(4) access to the airport on the return trip;

(5) in-flight return;

(6) egress return.

As the study was designed particularly to derive values for the Sydney Airport Study, stages (3) and (4) were not considered for international travel.

Since one important objective was to determine the distribution of time saving benefits between the employee, the employer, and the community, the passengers were divided into four types as follows:

(1) those travelling in their employer's time who were not specifically

compensated for travel UCP_{ER};

(2) those travelling in their employer's time who were compensated CP_{ER};

(3) those travelling in their own time who were not specifically compensated UCP_{EE};

(4) those travelling in their own time who were compensated CP_{EE}.

The proportion of passengers in category (2) was expected to be insignificant and was not considered further.

6.2.3. COMPONENTS OF THE VALUE OF TIME

The value of the time savings attributable to each passenger type on each journey stage was considered to be made up of four component parts:

(1) productivity effect;

(2) relative disutility cost;

(3) loss of leisure time;

(4) compensation.

The different values attributable to these elements and their different proportional contributions to the total caused the value of time savings to vary between journey stages and passenger types.

The productivity effect measured two things. First, the average value of an hour's working time, and second, the average value of work done during an hour's business travel. The average value of working time to the employer was estimated by summing the individual's salary, other non-salary benefits, directly attributable overhead costs, travel and accommodation expenses, together with the expected return to these expenditures, and this was then divided by the number of hours worked. Our estimate of the value of work done while travelling was estimated from the proportion of travel time spent working and the productivity of that work compared to equivalent 'ordinary' working time.

There is some doubt whether it is legitimate to make an allowance for work done while travelling in an employee's own time. Its inclusion could lead to the paradoxical situation where there was a net gain from deliberately increasing travel times outside the working day, merely to gain the advantage of the extra work done. If the value of such work exceeds the unit salary costs, the employee could be compensated at the average salary rate, with the surplus distributed between the employer and the community. On a practical level it can be claimed that some of the work done while travelling represents work that would be done in the employee's own time anyway. If this argument is accepted, then it is strictly not employee's but employer's time, and should be included in the total annual hours worked when calculating the hourly productive rate: the time taken in such work cannot be simply ignored, especially as the value of its output is presumably included in the annual total. Our preference was to consider work done while travelling in the employee's own time as an addition to the work that he would have done in his own time anyway, but the effect of adopting the alternative assumption is shown, where relevant.

The introduction of the relative disutility cost of travel was designed to bring into the valuation any costs or benefits of travel to the passenger himself. If businessmen prefer travelling to working under 'normal' conditions, then they gain some utility from travel, and conversely, if they prefer work under 'normal' conditions, they have a utility loss when travelling. This element of the value of travel time savings only applies to those travelling in their employer's time.

For travel which takes place in the employee's own time and is not compensated for by taking time off work later there is a loss of leisure time. Although there was uncertainty as to the appropriate value to put on the gains in leisure time associated with a reduction in travel time, the effort necessary to produce estimates of leisure time values better than those already available was beyond the scope of this study. It was assumed that the savings in travel time on each journey stage would be distributed between the employer's and employee's time in the same proportion as the total time for that stage. It was not considered that any possible redistribution of travel time between the two would be significant. (The assumption was nevertheless tested with a survey question relating to the response to an increase in travel time for a hypothetical journey. This indicated a similar distribution of time savings as for total travel time.)

Although compensation for travel is a transfer from the employer to the employee, it also helps to determine the allocation of the benefits attributable to time savings. Since it is one of the costs associated with the employment of a businessman who undertakes travel, it enters into the calculation of total annual costs when estimating the productivity effect and, since some compensation for travel is taxed, such payments are included in direct community costs.

6.2.4. FORMULATION OF THE VALUE OF TIME SAVINGS

The above elements were used to produce an estimate of the value of business travel time savings. The average value of an ordinary hour's work to the employer, V, was estimated from

$$V = AP_N^{ER} + OH_N^{ER} + TM_N^{ER} + P_N^{ER}$$

where AP_N^{ER} is the net average salary and salary-related cost per hour to the employer (including any payments in compensation for business travel in the employee's own time);

OH_N^{ER} is the net average overhead cost per hour to the employer;

TM_N^{ER} is the net travel and accommodation cost per hour associated with employee travel.

P_N^{ER} is the net contribution to employer's profit from an hour's employment.

Average hourly cost and profit rates were estimated from annual rates divided by the number of hours worked per year.

These basic quantities are expressed net of tax, since it is the net value which determines the benefit of any time savings to the employer. However, since the community also gains from an hour's employment (because of the contributions to taxation) these are included in the total value of an hour's saving in travel time and are equal to

$$CC = AP_{(G-N)}^{ER} + OH_{(G-N)}^{ER} + TM_{(G-N)}^{ER} + P_{(G-N)}^{ER} + CE_{(G-N)}^{EE}$$

where $AP_{(G-N)}^{ER}$, $OH_{(G-N)}^{ER}$, $TM_{(G-N)}^{ER}$ and $P_{(G-N)}^{ER}$ are the taxation revenues equivalent to AP_N^{ER}, OH_N^{ER}, TM_N^{ER} and P_N^{ER}, while

$CE_{(G-N)}^{EE}$ is the taxation revenue from travel time compensation payments to the business traveller.

The value to the employer of work performed during an hour's travel was found from

$$T = R.E.V$$

where T = value of work done while travelling;

R = proportion of travel time spent working;

E = proportional effectiveness of that time compared to ordinary working time.

Hence the over-all productivity effect (PE), associated with an hour's reduction in travel time, for a business passenger travelling in his employer's time was found from

$$PE = V - T$$

For business passengers travelling in their own time the over-all productivity effect was simply

$$PE = -T.$$

Now the total value of time savings on each journey stage (i) could be found from

$$VBTTS_i = \Sigma_j P_{ij} (PE_{ij} + L_{ij} \text{ (or } DC_{ij}) + CC_{ij})$$

where P_{ij} = the proportion of travel time on stage i that is of type
           j (that is CP^{EE}, UCP^{EE} or UCP^{ER});
         PE_{ij} = productivity effect for travel on stage i of type j;
         L_{ij} = the leisure value of time of stage i for type j;
         DC_{ij} = the relative disutility cost of stage i for type j;
         CC_{ij} = cost to the community of travel on stage i of type j.

The value of time for any combination of journey stages, such as the
two air stages for domestic travel, can be found by weighting the values
for the stages by the relative proportion of time spent on each of them.
The analyses for domestic and international business passengers were
kept separate.

The elements included in the estimate of *VBTTS* were all short-term
effects, with the possible exception of some work done while travelling.
In the long term it is possible that other costs might be involved. In
particular, either the employer or employee might incur removal costs
when an increase in travel time happens because of the relocation of an
airport. Where removal did occur, the net cost of removal plus new
travel costs was assumed to be lower than the travel costs would otherwise have been (or the move would not have been made). Although not
included in the general valuation, we did investigate, albeit rather super-
ficially, these possible longer-term costs.

6.3. EMPIRICAL FRAMEWORK

The investigation was limited to current passenger movements through
Sydney Airport and the interviews with passengers and employers
limited to those located in Sydney or Melbourne. It was assumed that
other Sydney Airport business passengers would have similar individual
(i.e. private) and employer-associated characteristics.

Two personal interview surveys were undertaken in Sydney and
Melbourne during September and October 1973. The first was an
employee survey on a sample of domestic and international business
travellers. They were selected by stratified random sampling from a
preliminary simple random sample taken at Sydney Airport. (Businessmen who had undertaken either, or both, domestic and international
journeys during the previous twelve months were included in the initial
sampling frame.) The second was an employer survey administered to
the accountant and/or personnel manager of various organizations
selected using random selection procedures.

The size of the employee sample was determined by the preliminary
sample size, the need for a desired distribution of employees in
accordance with two postulated important stratifiers (industry group
and frequency of travel), and time constraints imposed by the parent

study. It was decided to collect detailed information from a relatively small sample of business passengers rather than cover a larger sample less thoroughly. Domestic and international travel were separated in terms of distinguishing features such as journey length, time away from the office, and the seniority composition (and hence different income) of employees. Stratification was adopted as a means of using knowledge of the population as a whole to increase the precision of the sample. The industry grouping criterion refers to the division between governmental and private organizations (there was a reasonably uniform government policy on travel, in contrast to diverse private sector policies) and the breakdown of private enterprise by industry group in order to maintain reasonable sampling coverage of the business community. Frequency of travel was used as a stratifier on the assumption that the disutility of travel to the employee might be related to the frequency of travel and his ability to contribute towards productivity while travelling. The distribution of travellers by industry group was obtained from a more general survey of 50 000 air passengers carried out at Sydney Airport for the Project Team earlier in 1973. Of the 365 business passengers in the preliminary random sample 180 were included in the stratified domestic and international samples. Usable results were obtained from 89 domestic and 56 international passenger interviews, an effective response rate of over 80 per cent.

One hundred and seventy-two employers were initially selected for interview based on an expected response rate of 30 per cent. A telephone procedure was adopted to make initial contact with the employer to arrange an interview. The final sample (after allowing for non-response during the initial telephone contact or the initial interview) was 62, representing a response rate of 36 per cent.

The contents of the questionnaires are outlined in Tables 6.1 and 6.2. Both the form of the samples and the questions finally selected were based on a thorough pilot investigation. At an early stage attempts to match the employer sample with the employee samples were abandoned because of practical problems. For the main survey, interviews were conducted in Melbourne and Sydney during a six-week period. The details of the sampling, questionnaire design, and data collection procedures are described elsewhere (4).

6.4. THE EMPIRICAL RESULTS

6.4.1. PRODUCTIVITY EFFECT

A basic element in the determination of the value of business travel time savings is the value of the employee's output to the employer, expressed as an hourly rate (V). For this study the value is assumed to be equal to the net annual cost of employment to the employer plus the

TABLE 6.1   *Questionnaire content—employee surveys*

Group 1.	*General travel and user characteristics*
1.	Number of journeys in last 12 months
2.	Number of 1st class journeys in last 12 months
3.	Home or office origin/destination on last journey
4.	Commence and finishing times on last journey
5.	Access and egress modes on last journey
6.	Alternative acces and egress mode to private car usage
7.	Origin/destination of last journey
8.	Personal income group
9.	Position in the organization
10.	Industry group
11.	Business organization type (government versus non-government)
12.	Employment size of organization
13.	Road distance from home/office to airport
14.	Travel time to airport on last journey
Group 2.	*Work-specific characteristics*
1.	Weekly hours on which salary is based
2.	Additional number of weekly work hours
3.	Compensation for additional working hours
4.	Whether basic salary accounts for extra work hours
5.	Preference for additional work hours
6.	Income-related benefits —(type) only
Group 3.	*Travel time-specific characteristics*
1.	Percentage of travel time devoted to work
2.	Equivalent effectiveness between travel and office work circumstances.
3.	Time away from office on last journey
4.	Percentage of work during last journey able to be completed at office anyway
5.	Compensation for travel time
6.	Equivalent enjoyment between travel time and office time
Group 4.	*Other information*
1.	Access and egress mode costs claimed on last journey
2.	Influence of air travel on residential location

employee's contribution to profit, divided by the annual hours worked by the employee. Table 6.3 shows the items contributing to the cost of employment and their annual cost to the employer.

The first three items, the employee's salary, additional payments for extra hour's work, and the cost of income-related payments together constitute AP_N^{ER}, and were derived from the employee survey. Since these costs of employment are ordinary business expenses to the employer, with a corporate tax rate of 47·5 per cent, their net cost is only 52·5 per cent of the gross cost. The resultant annual net costs are thus $3976* for domestic and $4955 for international business passengers respectively.

* All dollar results are expressed in Australian currency

TABLE 6.2   *Questionnaire content—employer survey*

Group 1.	*General organizational and travel information*
1.	Number of employees in Australia and at the interview location
2.	Percentage of employees in Australia and at the interview location who undertake business air journeys
3.	Total number of employee domestic air journeys during the last month and 12 months
4.	Total number of employee international air journeys during the last 12 months
5.	Income distribution of business air travellers
6.	Official office working hours
7.	Road distance and travel time from office to airport
8.	Percentage of business air journeys on a 1st class ticket—domestic and international travel
9.	Distribution of business travel purposes
Group 2.	*Travel time-specific characteristics*
1.	Compensation for travel in employee's time
2.	Provision of a formal policy on compensation
3.	Percentage distribution of journeys beginning from office for various flight departure times—domestic and international travel
4.	Hypothetical time-cost trade-off
5.	Compensation for travel
6.	Time-adjustment period at end of an international journey
7.	Subsistence allowances and days per annum for which the allowance applies
Group 3.	*Work-specific characteristics*
1.	Income-related benefits—type and value
2.	Overhead costs—type and value
3.	Compensation for working extra time to complete scheduled work
4.	Contribution of traveller to profits
5.	Salary adjustments associated with attainment of an 'ideal' working environment
Group 4.	*Other information*
1.	Provision of car for access to airport
2.	Reinbursement for travel to airport—amount and policy
3.	Influence of air travel on office location

The average number of hours worked per year was found to be 2153 for domestic and 2106 for international business air passengers. Although we were unable to find any difference in the number of ordinary or additional hours of work per week for the two groups, we did find that the international group worked fewer weeks per year. Despite the emphasis on job completion, rather than working a set number of hours, a proportion of business passengers are compensated for work done (or for travel) performed outside ordinary working hours. The net cost of this compensation, which was made either in the form of additional payments or time off work (either officially or unofficially), was $299

TABLE 6.3   *The value of a business air passenger to his employer*

	Domestic business air passengers	International business air passengers
(a) Annual costs (A.$)		
Net salary cost to employer	5775	7350
Net cost of additional payments	299	409
Net cost of other employee benefits	1449	1679
Net overhead costs	1396	1478
Net travel and accommodation costs	710	1514
Total net cost of employment	9629	12 430
Employee's contribution to profit (at 12% of cost)	1155	1492
Total value of employee to employer	10 784	13 922
(b) Hours worked		
Average hours of 'ordinary' work per week	35	35
Average 'extra' hours work per week	12	12
Average weeks of work per year	45·8	44·8
Average hours of work per year	2152·6	2105·6
(c) Hourly value of employee to employer (A.$)	5·01	6·61

for domestic and \$409 for international business passengers, when averaged over the whole group. Other employee benefits cost \$1449 and \$1679 respectively.

The next item, the net hourly average overhead cost, was obtained from the survey of employers. The appropriate overhead costs were those directly attributable to the employees which would not have been incurred if that particular employee had not been employed. The items considered within this definition are given in Table 6.4, together with their annual cost. With many employers it was not possible to specifically isolate the overhead costs attributable to business air passengers, and in these cases we used the most appropriate costs available. It was not possible to differentiate between the costs attributable to domestic and international passengers. The annual costs in Table 6.4 were converted to net hourly costs by applying the same 52·5 per cent rate, and numbers of hours, as were used for converting income and income-related costs.

The net hourly travel and accommodation costs were also obtained from the survey of employers. The net costs were found to be \$710 and \$1514 respectively per year.

The net contribution that business air passengers made to profits (P_N^{ER}) was the most difficult component of V to evaluate. The survey of employers revealed a wide divergence of opinion on the 'profitability' of business air passengers in relation to other employees, as revealed by the following quotations:

". . . higher--the people who fly are our senior executives. They are the producers . . . These are the people who are making our place tick. If they were on the mean, we wouldn't be moving upwards. They would have to make an executive contribution at least higher than the mean of the division, otherwise they would not be in these positions . . . I believe that our profit figure would be 50 per cent lower without these people. They are responsible for doubling the profits."

". . . The basic answer is that we like to think everyone in the company pulls their weight. On the other hand we might say senior staff would make a greater contribution that junior staff and I would think the senior people would be most unhappy if I didn't say this--but of course the girls down below, I believe, are equally as productive and in fact we could not operate without them. While people's functions differ we would expect them to contribute equally to the company."

TABLE 6.4   *Employer's overhead costs*

	Gross cost per year (A.$) per business air passenger
1. Private secretarial staff attributable to the business air passenger	864·2
2. Office equipment	99·3
3. Office space	414·7
4. Recruitment, orientation, training	69·1
5. Payroll tax	612·2 (domestic) 767·0 (international)
6. Car parking space	200·0
7. Telephone rental and calls	200·0
8. Workers' conpensation insurance	200·0
Total	2659·5 (domestic) 2814·3 (international)

The empirical evidence suggested a mean rate of return on total expenditure attributable to the business air traveller of approximately 12 per cent per annum. The annual contribution to profit at this rate was calculated as $1155 for domestic travellers and $1492 for international travellers (i.e. 12 per cent of $AP_N^{ER} + OH_N^{ER} + TM_N^{ER}$).

Having estimated the value of an hour's ordinary working time to the employer (and to the community through the calculation of the equivalent taxation revenue) the over-all Productivity Effect itself could now be estimated, given values for R, the proportion of travel time spent working, and E, the proportional effectiveness of that time compared to ordinary working time. The product of the three quantities R, E, V gives the value to the employer of any work done while travelling (T).

The estimates of R and E were very subjective and were based on the employee's own estimates rather than those of their employers. Work was given a very broad definition which included discussions with colleagues, reading of journals, as well as preparing notes for meetings or preparing more official documents, and even thinking over work matters.

For travel in the employer's time (either UCP_{ER} or CP_{ER}) the Productivity Effect was the value of $(V - T)$, whereas for travel in the employee's own time it was simply $(-T)$.

The hourly values of the contribution to productivity during travel are given in Table 6.5. For domestic travel the outward egress trip has the highest productivity rate, followed by the outward in-flight trip stage. This could be explained on the grounds that the work undertaken is related directly to the activity at the outward destination, and that the work undertaken during travel is most efficiently done during travel because it is related to the purpose of the journey. The return access trip stage has the next highest productivity, perhaps because notes etc. made immediately after a meeting retain more of the information retained in discussion etc. than notes made on the next day or upon arrival back home or at work.

TABLE 6.5   *Value of work while travelling*

| | Proportion of time spent working | | Value to employer of work while travelling | |
	Domestic	International	Domestic A.\$ per hour	International A.\$ per hour
OUTWARD				
1. Access	·174	·096	0·54	0·33
2. In-flight	·448	·231	1·38	0·79
3. Egress	·481	—	1·48	—
RETURN				
4. Access	·329	—	1·01	—
5. In-flight	·270	·145	0·83	0·50
6. Egress	·270	·082	0·83	0·28

6.4.2. COMMUNITY BENEFIT

The net benefit which accrues directly to the community from savings in business travel time come from two sources. First, there is a gain in taxation revenue through the net increase in the output of goods and services (PE^{CC}) (the increase being net because of the loss of work done while travelling). This gain is partially offset by a loss from the second source, the reduced taxation revenue from payments to employees for compensation for travel in their own time (CE^{CC}).

With a corporate tax rate of 47·5 per cent the increase in taxation revenue from the first source will be 90·4 per cent (i.e. 47·5/52·5) of the value calculated for PE_N^{ER}. The losses in taxation revenue from reduced compensation payments will be 20·9 per cent and 19·0 per cent of the net value calculated for travel time compensation payments, based on the average tax rates of domestic and international passengers (international passengers have a lower tax rate despite their higher income, since they have more allowable expenses).

6.4.3. VALUE OF LEISURE TIME

The cost to the employee of travel during his own time is assumed equal to the leisure value of travel time (after allowance for the compensation effect when travel is compensated). The leisure values of time used were obtained from a 1971 work trip mode split study in Syndey (5) and comprise the opportunity cost (to the employee) of pure activity time, a 'comfort' differential, which is a function of changes in activity time, and the 'comfort' differential, which is a function of a constant amount of activity time. The latter is referred to as the disutility cost of travel. When travel occurs during the employer's time, then the only cost to the employee is the disutility cost of travel.

6.4.4. DISUTILITY COST OF TRAVEL

The disutility cost associated with business travel is a measure of the quality differential associated with travel, compared to spending equivalent time in the office. The amount of time is assumed constant, even though the relative disutility could change as the constant amount of relative time changes.

To place a shadow price on the relative disutility cost to the employee, a mechanism was devised to relate this to the known average salary rate. The value of the consumer surplus associated with working for a particular organization was obtained by asking:

Imagine yourself in an ideal working environment, doing the job you would want to do—but still occupying as many hours per week as your present job, and still involving the same extra benefits as your present job. What percentage, if any, below your present salary, would you be prepared to accept to do such a job?

This approach, based on the definition of an 'ideal environment', showed an average preparedness to accept a 14 per cent drop in salary. Because 'distant fields are greener' a 'present job approach' question was asked as a consistency check. This asked for the minimum salary an employee would be willing to accept for the job he was currently doing. This resulted in a mean reduction of 10 per cent in stated average salaries. Since the latter value related to a real situation, rather than the imaginary one implied by the former, the value of 10 per cent was used.

The disutility costs and leisure values of time for each trip stage are given in Table 6.6.

TABLE 6.6   *The disutility cost of travel and the value of leisure travel time* (A.$ per hour)

Trip stage	Disutility cost of travel		Value of leisure travel time	
	Domestic	International	Domestic	International
OUTWARD				
1. Access	1·03	1·31	0·86	1·18
2. In-flight	0·69	0·86	0·69	0·87
3. Egress	0·67	–	0·77	–
RETURN				
4. Access	0·57	–	0·82	–
5. In-flight	0·71	0·96	0·69	0·87
6. Egress	0·96	1·19	0·77	1·14

6.5. RESULTS AND CONCLUSIONS

The value of business travel time savings was obtained by inserting the quantities described in the preceding section into the equation given in section 6.2.4. Table 6.7 gives the elements of the VBTTS by trip stage and passenger type, while Table 6.8 shows the proportional distribution of passenger types. The final values for each trip stage are given in Table 6.9.

TABLE 6.8   *The proportional distribution of passenger types by trip stage*

Trip stage	Domestic Passenger type				International Passenger type			
	CP_{EE}	UCP_{ER}	UCP_{EE}	Total	CP_{EE}	UCP_{ER}	UCP_{EE}	Total
OUTWARD								
1. Access	0·1186	0·3186	0·5628	1·0000	0·1552	0·0653	0·7795	1·0000
2. In-flight	0·0507	0·4553	0·4940	1·0000	0·0827	0·0714	0·8459	1·0000
3. Egress	0·0940	0·4597	0·4463	1·0000	–	–	–	–
RETURN								
4. Access	0·0680	0·5831	0·3489	1·0000	–	–	–	–
5. In-flight	0·1130	0·3067	0·5803	1·0000	0·1335	0·1385	0·7280	1·0000
6. Egress	0·3396	0·2318	0·4286	1·0000	0·3624	0·1413	0·4963	1·0000

When weighted by the proportion of each type of passenger for each trip stage, the values of business travel time savings for domestic and international passengers are respectively 68·5 per cent and 30·3 per cent

TABLE 6.7 The value of business travel time savings, by trip stage and passenger type (A.$ per hour)

Trip stage	Domestic			International		
	CP_{EE}	UCP_{ER}	UCP_{EE}	CP_{EE}	UCP_{ER}	UCP_{EE}
OUTWARD						
1. Access	PE = 0·30–0·54 = –0·24 (0·30) CC = 0·66 (0·17) L = 0·86–0·49 = 0·37	PE = 5·03–0·54 = 4·49 (5·03) CC = 4·06 (4·55) DC = 0·63	PE = –0·54(0) CC = 0·49(0) L = 0·86	PE = 0·14–0·33 = –0·19 (0·14) CC = 0·38 (0·08) L = 1·20–0·23 = 0·97	PE = 6·51–0·33 = 6·18 (6·51) CC = 5·59 (5·89) DC = 0·81	PE = –0·33(0) CC = 0·30(0) L = 1·20
2. In-flight	PE = 0·16–1·38 = –1·22 (0·16) CC = 1·34 (0·09) L = 0·69–0·26 = 0·43	PE = 5·03–1·38 = 3·65 (5·03) CC = 3·30 (4·55) DC = 0·48	PE = –1·38(0) CC = 1·25(0) L = 0·69	PE = 0·08–0·79 = –0·71 (0·08) CC = 0·77 (0·05) L = 0·89–0·13 = 0·76	PE = 6·51–0·79 = 5·72 (6·51) CC = 5·18 (5·89) DC = 0·61	PE = –0·79 CC = 0·72(0) L = 0·89
3. Egress	PE = 0·30–1·48 = –1·18 (0·30) CC = 1·51 (0·17) L = 0·77–0·49 = 0·28	PE = 5·03–1·48 = 3·55 (5·03) CC = 3·21 (4·55) DC = 0·47	PE = –1·48(0) CC = 1·34(0) L = 0·77			
RETURN						
4. Access	PE = 0·28–1·01 = –0·73 (0·23) CC = 0·91 (0·16) L = 0·82–0·46 = 0·36	PE = 5·03–1·01 = 4·02 (5·03) CC = 3·80 (4·55) DC = 0·42	PE = –1·01(0) CC = 0·75(0) L = 0·82	PE = 0·13–0·50 = –0·37 (0·13) CC = 0·53 (0·08) L = 0·89–0·22 = 0·67	PE = 6·51–0·50 = 6·01 (6·51) CC = 5·43 (5·89) DC = 0·66	PE = –0·50(0) CC = 0·45(0) L = 0·89
5. In-flight	PE = 0·28–0·83 = –0·55 (0·28) CC = 1·07 (0·16) L = 0·69–0·46 = 0·23	PE = 5·03–0·83 = 4·20 (5·03) CC = 3·64 (4·55) DC = 0·49	PE = –0·83(0) CC = 0·91(0) L = 0·69			
6. Egress	PE = 0·77–0·83 = –0·06 (0·77) CC = 1·19 (0·44) L = 0·77–1·24 = –0·47	PE = 5·03–0·83 = 4·20 (5·03) CC = 3·80 (4·55) DC = 0·60	PE = –0·83(0) CC = 0·75(0) L = 0·77	PE = 0·36–0·28 = 0·08 (0·36) CC = 0·47 (0·22) L = 1·16–0·59 = 0·57	PE = 6·51–0·28 = 6·23 (6·51) CC = 5·64 (5·89) DC = 0·76	PE = –0·28(0) CC = 0·25(0) L = 1·16

Note: The figures in brackets are for the alternative assumption regarding the value of work while travelling

TABLE 6.9   *The value of business travel time savings (VBTTS) by trip stage*

Trip stage	Domestic A.$ per hour	Domestic % of gross salary per hour	International A.$ per hour	International % of gross salary per hour
OUTWARD				
1. Access	3·47 (3·84)	67·9 (75·1)	1·82 (1·98)	27·4 (29·8)
2. In-flight	3·69 (4·95)	72·2 (96·9)	1·58 (1·76)	23·8 (26·5)
3. Egress	3·41 (5·03)	66·7 (98·4)	–	–
RETURN				
4. Access	5·22 (6·17)	102·2 (120·7)	–	–
5. In-flight	2·91 (3·56)	56·9 (69·7)	2·40 (2·58)	36·1 (38·9)
6. Egress	2·51 (2·94)	49·1 (57·5)	2·75 (2·85)	41·4 (42·9)

of their average gross salary rates. When averaged over all business passengers the value is 62·9 per cent of the salary rate. The equivalent values when the work done while travelling is ignored are 85·1 per cent, 33·0 per cent, and 77·5 per cent.

The traditional assumption that savings in business travel time are equal to the passenger's wage rate was found to be inappropriate for business air passengers. The central value of 150 per cent used by the Commission on the Third London Airport is even less appropriate. The implications of the difference between domestic and international passengers is particularly significant where their interests might be in conflict, e.g. when deciding the division of services between complementary airports or when determining congestion priorities.

We have obtained clear evidence that business air travel does not all take place during normal work hours; the proportion for international business passengers is particularly low. Australian international business air passengers cannot be equated with their European counterparts. For the Australians the minimum flying time on an international trip is over 3 hours, while trips in excess of 12 hours are commonplace. In Europe a flying time of 3 hours is above average. Australian domestic air trips, with flying times which generally lie between half an hour and 2 hours, are comparable to European international trips. The characteristics of Australian domestic and international trips are similar to those in the United States.

The average values of in-flight time savings are 64·6 per cent of the average wage rate for domestic passengers and 30·0 per cent for international passengers. The corresponding values for surface stages are 71·7 per cent and 34·4 per cent. The value of savings in surface travel time is therefore at least 10 per cent higher than the value for in-flight time.

TABLE 6.10 *Comparison with the Roskill Commission figures*

| | Roskill (1969) | | | Sydney Study (1973) | | | | | | | | |
| | | | | Domestic passengers | | | International passengers | | | All passengers | | |
	£	% of income	% of total	A.$	% of income	% of total	A.$	% of income	% of total	A.$	% of income	% of total
Average income of business air passengers	3100	100·0	67·0	11 000	100·0	72·4	14 000	100·0	69·9	11 345	100·0	72·0
Income-related payments	362	11·7	7·8	1449	13·2	9·5	1679	12·0	8·4	1475	13·0	9·4
Overhead costs to employer	864	27·9	18·7	1396	12·7	9·2	1478	10·6	7·4	1405	12·4	8·9
Travel and subsistence	300	9·7	6·5	1352	12·3	8·9	2884	20·6	14·3	1533	13·5	9·7
Total	4626			15 197			20 041			15 758		

of their average gross hourly rates. When we involve all business passengers the value is 62·9 per cent of net salary. The equivalent values when the work done is regarded as utility are 85·1 per cent, 33·0 per cent, and 7·5 per cent.

The traditional assumption is that savings in air travel time are equal to the passenger's wage rate. We find this to be appropriate for business air passengers. The central value of net payment used by the Commission on the ... Appendix ... seems appropriate.

The implications of the difference between domestic and international passengers is particularly significant where their interests might be in conflict, e.g. when deciding on ... division of services between complementary airports or when determining operating frequencies.

We have obtained clear evidence that business travel does not all take place during normal working time ... especially for international business passengers is particularly low. Australian and international business air passengers cannot be compared directly with their counterparts. For the Australians the ... minimum flying time on an international trip is over 3 hours, while trips in excess of 12 hours are commonplace. In Europe a flying time of 3 hours is about average. Australian domestic air trips, with flying times which range ... between half an hour and 2 hours, are comparable to European short-haul air trips. The characteristics of Australian domestic ... air ... services are similar to those in the United States.

The average values of ... air travel time savings are 64·6 per cent of the average wage rate for domestic passengers and 30·0 per cent for international passengers ... The best ... saving values for surface stages are 71·7 per cent and 34·4 per cent. The value of savings in surface travel time is therefore at least 10 per cent higher than the value (z) in-flight time.

Since this value includes time spent in the airport terminals, the value of in-vehicle surface travel time is probably higher than the figures quoted.

We have already noted a significant difference in the total value of time savings compared with that given by Roskill (3). The relative composition of the annual gross employer costs of employing the air passenger are nevertheless similar in the two studies (Table 6.10). The average hours worked per year are also similar, the Australian figures being 8 per cent and 5 per cent higher than the Roskill results.

However, the main difference between the two studies lies in the way that the travel time savings in employee's time are dealt with relative to those in employer's time. If the Roskill assumption, i.e. that all savings were made in employer's time, had been applied to the Australian results, values of 160 per cent of the average wage for domestic and 179 per cent for international passengers would have been found. These results are

TABLE 6.11   *The incidence of the value of business travel time savings (%)*

Trip stage	Incidence of benefit	Domestic	International
OUTWARD			
1. Access	Employee	30·0	5·7
	Employer	22·6	59·6
	Community	47·4	34·7
2. In-flight	Employee	25·0	21·6
	Employer	16·0	55·5
	Community	59·0	66·1
3. Egress	Employee	22·2	—
	Employer	17·2	—
	Community	60·6	—
RETURN			
4. Access	Employee	40·0	—
	Employer	11·3	—
	Community	48·7	—
5. In-flight	Employee	18·1	11·4
	Employer	21·4	37·7
	Community	60·5	50·9
6. Egress	Employee	4·8	20·0
	Employer	21·6	40·3
	Community	73·6	39·7
Average Access/Egress	Employee	24·2	12·9
	Employer	18·2	50·0
	Community	57·6	37·2
Average In-flight	Employee	21·6	11·4
	Employer	18·7	37·7
	Community	59·8	50·9

very similar to the Roskill central estimate of 150 per cent.

A final part of the analysis relates to the distribution of travel time benefits between the passenger himself, his employer, and the community. All benefits eventually accrue to the community. This is implicit in the definition of resource costs. Here, however, we refer only to those benefits which, through increased taxation revenue, do not accrue specifically to the passenger or his employer. The proportion of benefits attributable to the community by this definition is a function of the taxation system and will vary between countries. The results shown in Table 6.11 are nevertheless likely to be typical for most developed countries.

REFERENCES

1 Hensher, D. A., *Consumer Preferences in Urban Trip-Making*, Commonwealth Bureau of Roads, Melbourne, Australia 1974 (5 volumes).

2 *Commission on the Third London Airport, Papers and Proceedings* vol. VII, Part I, H.M.S.O., 1970.

3 *Commission on the Third London Airport, Report*, H.M.S.O., 1971.

4 Hensher, D. A., Valuation of Business Travel Time Savings: A Study of Air Passengers, unpublished report prepared for R. Travers Morgan and Partners and the Sydney Airport Project Team.

5 Hensher, D. A., A Probabilistic Disaggregate Model of Binary Mode Choice, in D. A. Hensher (ed.), *Urban Travel Choice and Demand Modelling*, Australian Road Research Board Special Report no. 12, Melbourne, Australia; 1974.

About the Authors

Robin Carruthers has a degree in Economics and Political Institutions from the University of Keele. He is a consultant with R. Travers Morgan & Partners, for whom he was economic adviser with the Sydney Airport Project Team. Previously was a member of the Research Team of the Commission on the Third London Airport and aviation economist with the Department of Trade and Industry.

John H. Earp holds a degree in Civil Engineering. As Rees Jeffreys Road Fund Scholar he spent four years as a Research Associate at Newcastle University on research into problems of traffic generated from commercial centres. In 1967 he became a Senior Research Fellow and head of the Transportation Research Group at Southampton University. His interests lie in fields of transportation planning and traffic management and operations. Since September 1974 he has been an Associate with Freeman Fox and Associates.

Richard D. Hall Research Fellow, Transportation Research Group, Southampton University. Holds a degree in Mathematics with specialization in statistics and has been working within the Transportation Research Group since September 1969. Currently engaged on analysis of European inter-city trip generation.

Ian G. Heggie Director, Oxford University Transport Studies Unit. Holds degrees in Civil Engineering and in Politics, Philosophy, and Economics and has worked as a structural engineer (for Ove Arup & Partners), as well as an economic consultant (for the Economist Intelligence Unit). Is presently an economic adviser to Freeman, Fox & Associates.

David Alan Hensher Economist, Commonwelath Bureau of Roads (Australia). Holds a Ph.D. in Economics and has worked in university, consulting, and government environments over the last six years. A member of the U.S.A. Transportation Research Board Committee on Traveller Behaviour and Values, and a Visiting Fellow (1975) at the Transport Studies Unit, University of Oxford.

Margaret J. Heraty Associate, Alistair Dick & Associates. Previously Senior Public Transport Planner, Freeman Fox & Associates. Main interests lie in the field of public transport. Her career in transport planning consultancy began with a heavily mathematical bias but has come increasingly to concentrate on attitude research, particularly in relation to modal choice for both existing and new modes. Ms. Heraty read Mathematics at Imperial College and has an M.Sc. in Transport Studies from Cranfield.

Tony Jennings Lecturer in Economics, University of Leicester. Has carried out research on cost-benefit analysis and transport for the Department of the Environment, and undertook overseas assignments as Transport Advisor for the Commonwealth Secretariat. Author of a forthcoming book jointly with Clifford Sharp on Road, Rail, and the Environment. Currently researching on E.E.C. Common Transport Policy and Vehicle Taxation.

Michael McDonald is a Lecturer in the Department of Civil Engineering at the University of Southampton, specializing in Traffic Engineering. Prior to this appointment in 1971 he worked as an Experimental Officer within the Transportation Research Group at the same University.

Clifford Sharp Reader in Transport Economics, University of Leicester. Has carried out research on the allocation of goods traffic; urban passenger transport; and problems of transport and the environment. Author of the report 'Living with the Lorry', three books, and various articles on transport economics. Specialist adviser to the House of Commons Select Committee on Estimates (Trunk Roads and Motorways) and Nationalized Industries (National Freight Corporation).

Index